SAT Math in the Classroom

Bring SAT Math prep into the classroom to enhance student learning! In this new copublication from Routledge and test-prep experts A-List Education, you'll learn how the updated SAT exam is closely aligned with the Common Core, making it easy to weave test prep into your curriculum and help students hone the skills they need for college readiness. The book is filled with practical examples of how the Common Core State Standards are connected to specific sections, question types, and strategies applicable to the SAT, so you can simultaneously prepare your students for the test while improving their overall math and reasoning skills.

Bonus: A Study Guide to help you use the book for school-wide professional development is available as a free eResource download from our website: www.routledge.com/9781138668362.

A-List Education is an educational services provider serving more than 50,000 students with tutoring programs across the U.S. as well as in the UK, Dubai, Switzerland, and China.

Other Books Available From Routledge and A-List Education

(www.routledge.com/eyeoneducation)

SAT ELA in the Classroom:
Integrating Assessments, Standards, and Instruction

ACT ELA in the Classroom:
Integrating Assessments, Standards, and Instruction

ACT Math in the Classroom:
Integrating Assessments, Standards, and Instruction

SAT Math in the Classroom

Integrating Assessments, Standards, and Instruction

A-List Education

Routledge
Taylor & Francis Group
NEW YORK AND LONDON

A LIST

First published 2017
by Routledge
605 Third Avenue, New York, NY 10017

and by Routledge
2 Park Square, Milton Park, Abingdon, Oxon, OX14 4RN

Routledge is an imprint of the Taylor & Francis Group, an informa business

Library of Congress Cataloging-in-Publication Data
Names: A-List Education.
Title: SAT math in the classroom : integrating assessments, standards, and instruction / by A-List Education.
Description: New York : Routledge, 2017. | Includes bibliographical references.
Identifiers: LCCN 2016009872 | ISBN 9781138668355 (hardback) | ISBN 9781138668362 (pbk.) | ISBN 9781315618678 (e-book)
Subjects: LCSH: Mathematics—Examinations, questions, etc. | Mathematics—Study and teaching. | SAT (Educational test)
Classification: LCC QA43 .S2658 2017 | DDC 510.71/2—dc23
LC record available at https://lccn.loc.gov/2016009872

ISBN: 978-1-138-66835-5 (hbk)
ISBN: 978-1-138-66836-2 (pbk)
ISBN: 978-1-315-61867-8 (ebk)

Typeset in Palatino
by Apex CoVantage, LLC

Contents

eResources vii
About the Author ix
Introduction and Overview xi

1 **About the SAT** 1
Format 3
Math 5
Changes to the Test 9

2 **Alignment With Common Core Math Standards** 11
How to Read the Math Standards 11
Alignment 13
How to Read This Section 15
Alignment Beyond the Test 17
Standards for Mathematical Practice [MP] 18
Number and Quantity [N] 20
Algebra [A] 23
Functions [F] 27
Geometry [G] 30
Statistics and Probability [S] 34

3 **SAT Math in the Classroom** 37
Overview 37
Timed Practice 41
Problem-a-Day 47
Translation Exercises 50

Appendix: All Math Alignment Tables 67
Additional Resources 93

eResources

This book is accompanied by free online eResources, including a Study Guide to help you work on this book with colleagues, as well as additional materials to help you with school-wide implementation of the ideas in this book. To access the eResources, go to www.routledge.com/9781138668362 and click on the eResources tab. Then click on the items you'd like to view. They will begin downloading to your computer.

About the Author

A-List Education was founded in 2005 with a mission to bring innovation and opportunity to education, empowering students to reach their true potential. We work with schools, school districts, families, and nonprofits and provide tailored solutions for specific learning and curriculum needs—ultimately working to improve college readiness and access. Our staff comprises experienced and passionate educators, each with a distinctive and personal approach to academic success, and our management team collectively possesses more than 75 years of tutoring, teaching, and test preparation experience. We now provide leading-edge education services and products to more than 500 high schools and nonprofit organizations, helping more than 70,000 students a year in the United States and around the world.

A-List has a variety of offerings for SAT and ACT preparation, including:

- **Textbooks** for students studying individually or for teachers conducting classes. Our content not only emphasizes test-taking techniques but also reinforces core skills, which empower students for academic success long after taking the test.
- **Professional development** to help schools and organizations set up their own courses. Our seminars create valuable educational expertise that will allow teachers in your district to bring content and problem-solving strategies directly into their classrooms.
- **Direct course instruction** using our own staff. Our dedicated and experienced teachers receive intensive training before entering the classroom and have proven track records of empowering students to reach their academic potential.
- An **online portal** to remotely grade practice tests and provide supplemental material. This platform removes the burden of grading complex tests without requiring customized technology and provides supplemental material for your ongoing courses.

- Individual one-on-one **tutoring**. Our instructors help students deliver average improvements of more than three times the national average in the United States. In addition, our students routinely gain acceptance to their top choice schools and have been awarded millions of dollars in scholarships.

Visit us at **www.alisteducation.com** to learn more.

Introduction and Overview

The SAT and the Common Core

The Common Core State Standards Initiative (CCSSI) is a program designed to unify the state standards across the United States so that students, parents, and educators have a clear sense of what skills students must acquire in school to become ready for college or a career. It is an independent initiative in which states may voluntarily choose to participate, but by the end of 2015, 42 states plus the District of Columbia had chosen to adopt the standards. The standards are designed to be rigorous, clear, and consistent, and they are based on real evidence to align with the knowledge and skill necessary for life beyond high school.

The program is controversial to be sure, but a strong foundation is already in place, and schools around the country are working to align their own materials and programs with the newly adopted standards. The standards specifically do not outline exact requirements for curriculum, such that schools and districts can still make their own choices about how to run their classes while still adhering to the standards. As a result, some schools may struggle to find the right material.

However, there already exists a comprehensive source of material that addresses the wide range of skills and knowledge that the Common Core emphasizes: the SAT.

Not all schools currently offer preparation programs for the SAT, but even those that do tend to treat the test as distinct from normal schoolwork. The test is seen as supplementary, and preparation is an unpleasant game in which students learn tricks to game the system without actually learning skills. This view of the test, however, is not just uncharitable but false. In fact, SAT preparation can fill many of the gaps to help schools align their curricula with the Common Core.

The SAT requires students to use many of the same math and reading skills that are the goals of the standards. It is designed to identify whether students are ready for college by testing them on the skills and knowledge they will need when they get there. In fact, the designer of the SAT, the College Board, was actively involved in the creation of the standards, and their own research about college readiness was integral to the program. Furthermore, recent changes to the test in 2016 were initiated in no small part to bring the test further in alignment with the standards.

Preparing for the test can accomplish two goals at once. Test preparation's primary goal is to prepare for the test itself, helping students maximize their scores on the test and thus improving their chances of being admitted to the colleges of their choice. Beyond pure admissions, students' test scores can have a number of uses for different programs and institutions. For example,

- **College admissions**. Roughly half of a student's admissions profile is composed of a combination of GPA and SAT or ACT score. A high score can be a huge differentiator for the majority of elite universities and a minimum hurdle for the majority of state universities.
- **Scholarships**. There are billions of dollars of aid in private and school-based scholarship money tied directly to test scores.
- **Community colleges**. Even at community colleges with low graduation rates, good scores can allow students to avoid placement in remedial classes.
- **Military**. For students interested in the military, baseline scores can qualify a student for officer training as opposed to regular enlistment.
- **State assessment**. The SAT is increasingly used as a statewide student assessment to identify achievement of particular benchmarks.

But test preparation is also a valuable activity in itself; students will also be working on honing and strengthening the skills they need for college readiness. Furthermore, the test material itself is valuable even beyond their application to the actual test. The passages, essay prompts, and mathematical concepts contained herein can be divorced from the SAT. You do not have to actually take the SAT to draw value from reading and analyzing its passages, analyzing its grammatical structures, or attempting its math problems.

It is for these reasons that some states have decided to use the SAT as its primary measurement for high school achievement, rather than a more explicitly standards-based assessment.

Alignment

In 2010, the College Board produced an alignment study to show how the standards align with the skills that the SAT assessed. The results of these studies show that the test significantly aligned with the standards.[1]

Since then, the test was radically changed, starting with the PSAT in 2015 and the SAT in 2016. While the changes to the test were clearly motivated by alignment to the standards, the College Board has not yet produced an official alignment study for the new SAT. They have not officially said much to date about the Common Core specifically. In a guide to implementing the new test, in response to alignment with the CCSS, the document states: "The redesigned SAT measures the skills and knowledge that evidence shows are essential for college and career success. It is not aligned to any single set of standards." This seems to distance the test from the notion that the Common Core was the motivation for the redesign. But it also notes that the skills for the new SAT draw from the same evidence base as "state academic standards including the Common Core, the Texas Essential Knowledge and Skills, and the Virginia Standards for Learning—as well as in the best college-prep curricula."[2]

However, we at A-List want to take a closer look at how exactly the new test aligns with the Common Core. We have compared the old alignment document and the new test material to project the alignment for the new test.

What does this alignment tell us? Several things:

1. The redesigned test aligns strongly with the CCSS.
2. The redesigned test does not align *perfectly* with the CCSS. There are a number of standards that are not relevant to the SAT.
3. While some changes to the SAT have made it more aligned with the CCSS, others have made it less so. As a whole, the test isn't necessarily more aligned with the test than it already was.
4. *But the old SAT was already strongly aligned with the CCSS.*

The College Board's study demonstrated that the skills tested by the SAT align with the skills of the standards. This study is heavily researched and incredibly thorough. However, it has not been updated for the new test and it does not link these skills to real test material. This book will use that information to demonstrate specifically the connection between the test and the standards in a practical manner. Furthermore, this book will show that test material can be adapted and expanded to address even those standards that are outside the scope of the test.

About This Book

This book has two main goals:

1. To show how specific SAT Math material aligns with the Math CCSS.
2. To discuss how to incorporate SAT preparation into your regular math classes outside of an explicit test preparation class.

Why do we focus on the CCSS? First of all, because the grand majority of states use it. There's no shortage of debate about the value of the standards, but it's undeniable that they are in place throughout most of the country. Even some states that are moving away from the standards are doing so more in name than in practice and are keeping the content of the CCSS in place.

Of course, not every state uses CCSS, but it remains a useful framework for connecting SAT material to classroom material. We can't run through every state's particulars in one book, but the popularity of the CCSS make them a convenient reference. If you don't use them, you can still use CCSS as a touchstone to compare to your own state standards. Even if you do use CCSS, your state may also have its own assessments or graduation requirements that deviate from CCSS. The alignment information is one piece of the picture.

Second, some states are actually using the SAT as their main statewide assessment. They are generally doing so because of the SAT's alignment to the Common Core. Is this a good idea? That's debatable, and we won't take sides here. The question is part political, part pedagogical, often emotional. However, we can offer our research and expertise in the test so you can see for yourself where it coincides with your curriculum and where it doesn't.

Structure of the Book

Chapter 1 will describe the structure and content of the SAT's Math sections for those who may be unfamiliar with the test, along with discussion of how the test has changed and some effective test-taking techniques. Keep in mind that this is a general overview. It draws information from our main textbook, *The Book of Knowledge*, which is the product of years of experience with the test, and goes into much greater detail about the test's content and the most effective strategies.

Chapter 2 will connect SAT Math to the Common Core math standards. This section will take a closer look to show where the test does and does not align with these standards. It will go through each content category and domain to discuss what specific sections, question types, or strategies align with the standards in question. Additional discussion also describes how, even when the test does not align with the standard, test material can be pushed beyond its intended scope in order to do so.

Chapter 3 will focus on how to use all this material in the classroom. This could mean using your math classes as explicit preparation for the test; it could mean using test material to supplement your regular classes; it could mean preparing for the test as a tool with which to get your students to meet the standards; or it could be a combination of these things.

The Appendix will list all of the complete alignment tables discussed in Chapter 2, followed by a bibliography and suggestions for further reading.

Notes

1. Interestingly, this document is no longer available on the College Board's website, possibly to avoid confusion between the old test and the new test. It is still possible to find it online hosted elsewhere with a bit of searching.
2. *College Board Guide to Implementing the Redesigned SAT*, October 2014, https://collegereadiness.collegeboard.org/pdf/college-board-guide-implementing-redesigned-sat-installment-2.pdf

1

About the SAT

The SAT is a college admissions test, first and foremost. It was designed for that purpose and has been used as such since 1926. In the century since its creation, it has substantially changed a number of times, but its primary goal has remained the same: to provide a standardized metric for colleges to be able to judge students from disparate backgrounds.

With the redesign of the SAT in 2016, however, the College Board has been trying to shift the focus of the test more towards the K–12 market. Much of the rhetoric surrounding the redesign centered around "real-world skills" and "math that matters most", eliminating "irrelevant vocabulary".

Of course, this is not the first time the SAT has been redesigned—most recently in 2005 when analogies were removed and the Writing component and essay were added, or in 1994 when calculators were first permitted and non-multiple-choice grid-in questions were added. It seems like every redesign has been surrounded by rhetoric of focusing on real-world, important skills. With this redesign, the College Board has made an active effort to emphasize college readiness benchmarks to a greater degree than they did before. They are reorganizing their suite of tests, including versions the PSAT starting as early as eighth grade, and moving toward a longitudinal assessment of benchmarks throughout high school. They are moving past the market of individual students taking the test for college and marketing their tests directly to schools, to be used both for college admissions and for skills assessment.

Is this a good idea? Who knows? Very limited data is currently available, so it's impossible to say yet whether the test does a good job at assessing these skills. We don't even know how well the new test serves for admissions. The old SAT, when combined with high school GPA, was a stronger predictor of college performance than was either of those metrics alone. We shall see how the new tests shape up when we see more data and that new-test smell starts to wear off.

In the meantime, it's here, so we must acknowledge it and deal with it. While we don't know much about how the scores will shape up, we do know quite a bit about what's on the test, enough that we can make a concerted effort to prepare students for it.

We want to help you incorporate SAT material into your classroom in order to prepare students, without running an explicit SAT prep course. To do so, the first and most important thing is simply to know what's on the test and what the test is like. The best way to do that is firsthand: *you should absolutely do some official practice tests yourself.* There are four full-length SATs and one PSAT available for free download on the College Board's website. Go do one and see what you think.

In the meantime, we're not going to go through our whole prep book (it is, however, for sale on our website!), but we do want to give you an overview of the structure and content of the test.

Format

The SAT is split into two subjects, **Evidence-Based Reading and Writing** and **Mathematics**. The test has four sections, two for each of the subjects, plus an optional fifth section for the **Essay**. The test lasts 3 hours, or 3 hours and 50 minutes if you choose to do the essay.

Besides the Essay, most sections have only multiple-choice questions, with one exception: each Math section will contain some questions for which students must produce their own responses. All multiple-choice questions will have four possible choices.

The two sections for Evidence-Based Reading and Writing (henceforth called EBRAW) will cover two different content areas: **Reading** and **Writing and Language**. The two Mathematics sections will cover mostly the same material, but on one section calculators are permitted and on the other they are not.

Table 1.1 SAT Format

Section	Portion	Number of Questions	Time	Description
Evidence-Based Reading and Writing	1. Reading Test	52 questions	65 min	5 passages, each with 10–11 questions on reading comprehension
	2. Writing & Language Test	44 questions	35 min	4 passages, each with 11 questions on grammar, usage, and style
	Total	**96 questions**	**100 min**	
Math	3. No Calculator	20 questions	25 min	15 multiple-choice questions 5 student-produced response questions (grid-ins)
	4. Calculator OK	38 questions	55 min	30 multiple-choice questions 8 student-produced response questions (grid-ins)
	Total	**58 questions**	**80 min**	
Essay	*5. Essay*	*1 essay*	*50 min*	*Optional. One 1–4 page essay.*
	Total	**154 questions**	**3 hours**	
	with essay	*154 questions + 1 essay*	*3 hours 50 min*	

Scores

Each student will get a *Section Score* ranging from 200 to 800 for each of the two subjects. These scores are added together to produce a *total score* ranging from 400 to 1600. These are scaled scores, calculated by taking the number of right answers and converting them to the scaled score using a scoring table unique to the particular test the student took. This is done to ensure that differences in difficulty across forms will not affect scores. Note that students do not lose points for wrong answers.

Each student will also get three *Test Scores* in Reading, Writing and Language, and Mathematics ranging from 10 to 40. These scores are tied directly to the section scores: the EBRAW section score is the sum of the Reading and Writing test scores multiplied by 10. The Math section score is your Math test score multiplied by 20.

Each student will also get a variety of *subscores* for different types of questions within each section. These scores are either on a 1 to 15 or 10 to 40 scale. Students who take the essay will get three scores for the essay, each on a 2 to 8 scale. The *essay scores* are separate from the rest of the test; they will not be factored into students' EBRAW section scores or any other scores.

This book will focus on the Mathematics portions of the test. We will not be discussing the Evidence-Based Reading and Writing section or the Essay.

Table 1.2 Math Subscores

Name	Type	Scale	Sections
Heart of Algebra	Subscore	1–15	Math
Passport to Advanced Math	Subscore	1–15	Math
Problem Solving and Data Analysis	Subscore	1–15	Math
Analysis in Science	Cross-Test Score	10–40	Reading, Writing and Language, Math
Analysis in History/ Social Studies	Cross-Test Score	10–40	Reading, Writing and Language, Math

Math

Format

Table 1.3 shows a general overview of the sections:

Table 1.3 Math Format

Section	Time	Content
Section 3: No calculator	25 minutes	20 questions: 15 multiple-choice questions 5 grid-ins
Section 4: Calculator OK	55 minutes	38 questions: 30 multiple-choice questions 8 grid-ins
Total	**80 minutes**	**58 Questions**

Each section contains some four-choice multiple-choice questions followed by non-multiple-choice "Student-Produced Response Questions" (which we call "grid-ins" for short). For these questions, students must determine their own solution and fill it into a grid on their answer sheet. Answers may contain no more than four digits (including nonnumerical characters like a decimal point or fraction slash). No partial credit for methodology is awarded.

All questions are ordered by difficulty. Multiple-choice and grid-in questions are numbered continuously within a section but are ordered by difficulty separately. That is, in section 4, question #30 will be a hard multiple-choice question, while question #31 will be an easy grid-in.

Calculators

Calculators, including most types of graphing calculators, are permitted on Math section 4 only. Calculators of any kind are not permitted on section 3. This was one of the biggest changes to the Math test; previously, calculators were allowed on all sections. (Of course, some of us remember a time before 1994 when calculators weren't allowed on *any* questions. Some of us are very old.)

What does this really mean, though? What's the difference between the calculator questions and the no-calculator questions?

Honestly, it means *very little*. Some students may get anxious about calculator permission. They worry about not having it on section 3 and then overuse it on section 4. Most questions between the two sections are incredibly similar. If we showed you a question at random, you probably wouldn't be able to tell which section it's from. For many if not most no-calculator questions, you wouldn't really use your calculator even if you had one. And you shouldn't need your calculator on *every* question where it's allowed.

True, there are some questions on the no-calculator section that you may wish you had a calculator for, but they're rare. The test isn't going to force you to multiply two five-digit numbers by hand. Similarly, you will certainly see a few questions on the calculator section that do involve tough computations that you don't want to do by hand, but those are the exception, not the rule. Most of section 4 is *calculator optional.*

SAT Math Techniques

While much of the test comes down to straightforward knowledge of math, the test isn't exactly like the kind of tests students see in school. A-List's SAT book, *The Book of Knowledge*, has a number of tips and strategies for the special kinds of issues that arise on the SAT.

Error Avoidance

One of the biggest obstacles students face is *distractor choices*. For the most part, the wrong answers on the question aren't random. They're carefully selected to be tempting for students. There are specific reasons why a student may reasonably believe them to be correct. It could be because of a careless math mistake, like not distributing across parentheses or adding instead of subtracting. It could be because they didn't read the question carefully, like solving for x when the question asks for y.

Techniques

In general, SAT math is not about knowing a lot of content as much it is about knowing a limited amount of content *very well*. The number of rules students need is small compared with what you actually teach in school, but these rules will be combined in unexpected ways. Therefore, learning multiple methods of problem solving is a crucial part of preparing for the test.

For example, often a question will appear to require complex algebra, but the problem can be solved with simple arithmetic. This can be done by choosing values for the variables (we call this **Plug In**) or by testing the values in the choices (we call this **Backsolve**). It takes practice to get comfortable using methods beyond what you're used to, but these methods are crucial to score improvement. If your students don't change their methods, they won't change their scores.

Content

Heart of Algebra

These questions are the most common type (though the main three types are fairly evenly distributed) and have the second dumbest

Table 1.4 Math Content Areas

Content Category	Section 3 No Calculator	Section 4 Calculator OK	Total	Percent of Test
Heart of Algebra	8	11	19	33%
Passport to Advanced Math	9	7	16	28%
Problem Solving and Data Analysis	0	17	17	29%
Additional Topics in Math *(no subscore)*	3	3	6	10%

name. The main theme of these questions is **algebra of linear equations**. These are simple expressions that have no exponents and can be represented as a straight line in a graph. Major topics include:

- Writing, solving, and understanding linear equations.
- Writing, solving, and understanding linear inequalities.
- Graphing lines.
- Solving systems of two linear equations.

Passport to Advanced Math

These questions involve **higher-order algebra** and have the dumbest name. Major topics include:

- Using and understanding function notation.
- Manipulating polynomials: simplifying, factoring, multiplying, and dividing.
- Writing, solving, and understanding quadratic equations (including using the quadratic formula).
- Writing, solving, and understanding exponential functions.
- Graphing parabolas and higher-order functions.

Problem Solving and Data Analysis

Note that these questions *only* appear on section 4, the calculator section. Questions in this content area fall into two general content areas:

- *Proportions*: fractions, ratios, unit conversions, percentages.
- *Statistics*: data visualization, tables, scatterplots, probability, averages, sampling.

Additional Topics in Math

This is the only category that doesn't give you a subscore, and it's by far the smallest, with only six total questions on the test. These questions fall into three general content areas:

- *Geometry:* angles, triangles, circles, area, volume.
- *Trigonometry.*
- *Imaginary and complex numbers.*

Changes to the Test

The SAT went through a well-publicized redesign, beginning with the PSAT in October 2015 and the SAT in March 2016. Ultimately, since the pre-2016 test is dead, it doesn't matter what the changes were; all that matters is what the test looks like now. But it's worth taking a quick look at those changes because (a) you may be more familiar with the older version of the test, and (b) examining what the College Board changed can help us understand their thinking about what they want the test to be.

Test Structure

- **Fewer sections.** The old test had ten sections, three in each subject plus a variable equating section in any subject. Each section was 25 minutes or less. The new test has fewer sections and more time per section.
- **Fixed section order.** The old test would mix up the order of the sections on each test administration so students wouldn't necessarily know what was coming next. The new test uses the same section order each time.
- **No equating section.** The old test contained one section that was used for setting the scoring table and did not count towards the student's score. The new test does not contain such a section.
- **No guessing penalty.** The old test subtracted a fraction of a point for each wrong answer in order to counteract the benefit of random guessing. The new test treats incorrect answers the same as blanks.
- **Fewer choices.** The old test had five choices for all multiple-choice questions. The new test has four.

Scoring

- **Two main scores.** The old test had three main scores, Reading, Math, and Writing, each scored from 200 to 800. The new test has two such scores, combining Reading and Writing into one 200–800 score.
- **More subscores.** The old test had (with the exception of Writing) no scores other than the main 200–800 scores.

The new test has many additional subscores for all parts of the test. A student's score report for the new test will contain 18 distinct scores in all.

Math Test

- **No-calculator section.** The old test permitted calculators in all math sections. The new test permits it only on one of the two sections. In addition, the old test was truly calculator optional, such that each question could be done by hand with minimal effort. The new test features some questions which, while *technically* could be done by hand if you're brave enough, really do require a calculator.
- **More grid-ins.** The old test only had 10 grid-in questions out of 54 questions (19%), while the new test has 13 out of 58 (22%). Additionally, the old test had grid-ins on only one section out of three, while the new has them on both sections.
- **Refocused content.** The old SAT tested algebra, arithmetic, geometry, and statistics in roughly equal amounts, with somewhat more attention on algebra and less on statistics (roughly 35%, 30%, 25%, and 10%, respectively). The new test has virtually banished geometry to less than 10% of the test and added more statistics questions and *much* more algebra—Algebra 1 and 2 together make up over 60% of the new test.
- **More content.** In addition to refocusing the content, the new SAT features some added more difficult concepts that were either minimally tested or entirely absent from the old test. These include exponential growth, dividing polynomials with remainders, trigonometry, and complex numbers, to name a few.

2

Alignment With Common Core Math Standards

How to Read the Math Standards

There are two components to the mathematics standards.

First, the Standards for Mathematical Practice describe the skills necessary for effective mathematical thinking and proficiency. Rather than specific content, these standards refer to habits and practices that students should develop and utilize throughout all their mathematics courses, from kindergarten through twelfth grade.

Second, the grade-level standards describe what students should be learning for every grade level they achieve throughout their schooling. Since we are concerned with the SAT, we will focus solely on the standards for high school. Unlike the ELA standards, the mathematics standards are not broken up into individual grades for high school; there is one set of mathematics standards for all of high school.

These standards are divided into several conceptual categories:

- Number and Quantity
- Algebra
- Functions
- Geometry
- Statistics and Probability
- Modeling

These roughly correspond with individual courses that students would take in high school, but they do not have to be so discrete.

Standards from one category may well also apply to work in courses that focus on other categories.

Each of the standards within the categories is further arranged into different groupings. Standards are arranged into clusters of closely related concepts. Closely related clusters are grouped into domains. Domains are grouped into conceptual categories. Note that standards are numbered continuously within a domain. Numbering does not restart with a new cluster, only with a new domain. Furthermore, some standards may be subdivided into two or more subskills that are related to that standard.

Each standard has a code containing three parts: a letter denoting the conceptual category, a series of letters denoting the domain, and a number denoting the standard. If the standard has any subskills, the skills will be denoted by a letter following the standard number. For example, "F-IF.7c" denotes:

- the Functions category [F],
- the Interpreting Functions domain [IF],
- the seventh standard within that domain [7], and
- the third subskill associated with that standard [c].

Code: [category]-[domain].[standard][subskill]

Modeling is slightly different from the other conceptual categories listed above. The standards in this category have less to do with specific content and more to do with applying math content and skills to real-life situations. Skills like these are integral to the SAT, and questions that ask students to determine how to apply mathematical concepts to actual settings to solve specific problems are prevalent. Rather than producing a separate list of modeling standards, these modeling standards are distributed throughout the other conceptual categories. In all alignment tables, any standard that is also a modeling standard will be marked with an asterisk (*).

Finally, there will be some standards shown here that are more **advanced** than others within their category. The goal of the Common Core State Standards is to ensure that students are ready for college and future careers, and the standards were specifically written with that in mind. However, within these categories there are also some standards aligning to skills that are not necessary for college and career readiness but may be necessary for more advanced mathematics courses, such as calculus. Standards that go beyond college and career readiness are marked with (+).

Alignment

How Was Alignment Determined?

We used a number of sources to determine alignment between the redesigned SAT and the Common Core State Standards:

- We started with the College Board's own study aligning the old SAT to the CCSS. We made adjustments according to how the test has changed or not changed.
- The College Board's "Test Specifications for the Redesigned SAT" gives descriptions of the new test's content and skills, as well as background on how the test was developed.
- Questions directly from the test maker are always the most reliable guide to what the test will look like, but unfortunately there isn't much material available. As of this writing, four full-length practice SATs and one PSAT were released on the College Board's website. The same four SATs are also published in their book *The Official SAT Study Guide.*
- Khan Academy (www.khanacademy.org) publishes drills and problem sets produced in conjunction with the College Board. Since these questions have the test maker's official approval, they provide a larger data set for the kinds of questions that will be on the test.

The College Board Alignment Study

In 2010, the College Board produced an alignment study to show how the standards aligned with the skills assessed by the old-format SAT. The results showed that the test significantly aligned with the standards.

Since then, the SAT has changed, of course, and the College Board has yet to produce a new alignment report. In order to determine alignment with the redesigned test, we used the initial alignment report and updated it according to what we know about the changes to the test. This includes the information the College Board has provided about the reasoning behind the redesign and specifications for the new test, along with the actual practice tests that have been released. The alignments provided here, then, are our own interpretation of how well the standards aligned, based on information and data from the College Board.

The following chart outlines the results for Math standards, showing what percent of the standards in each strand align with each test's stated skill set.

Table 2.1 Alignment Summary

Common Core State Standard Category	Old SAT Alignment	Redesigned SAT Alignment
Standards for Mathematical Practice [MP]	**100%**	**100%**
Standards for Mathematical Content, Grades 9–12	**78%**	**83%**
Number and Quantity [N]	27%	40%
Algebra [A]	94%	94%
Functions [F]	75%	85%
Geometry [G]	91%	93%
Statistics and Probability [S]	94%	94%
Modeling (*)	93%	93%
Advanced (+)	43%	49%
College and Career Ready (not (+))	91%	97%

The actual alignments performed were for a larger suite of materials than just the SAT. The company offers additional tests and programs besides the SAT, such as PSAT/NMSQT. Each component was aligned against the standards for the grade level to which the particular program is meant to be given. This document will focus on alignment for the SAT test itself, which was aligned to the Standards for Mathematical Practice and the Standards for Mathematical Content for high school.

Some standards show what we call "partial alignment". In the SAT alignment report, certain standards that were shown as aligned had additional comments qualifying that alignment. The table shows percentages of standards that show any alignment; it does not distinguish partial and complete alignment.

How to Read This Section

The following discussions will be grouped by conceptual categories shown above. At the start of each chapter, we will list all of the domains and clusters within that category. For each cluster, the number of standards within that cluster is given in parentheses. The number of subskills associated with those standards is also given, where applicable.

Figure 2.1 Sample Headings

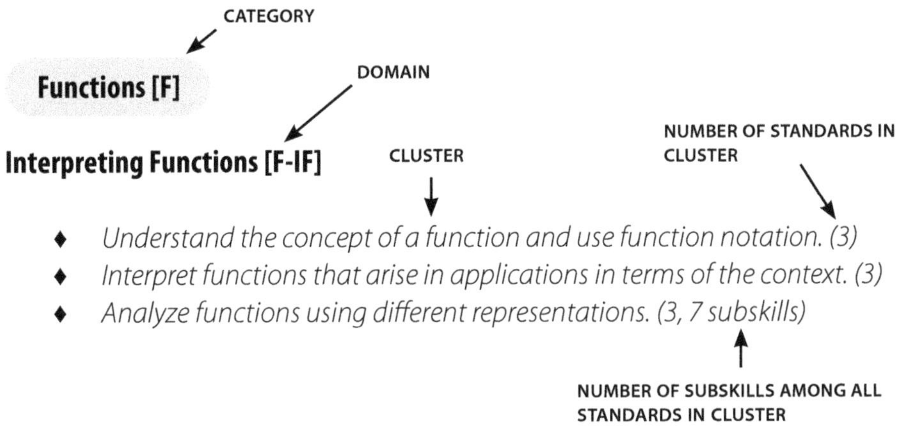

For the sake of clarity, we do not list all individual standards in the discussion chapters. All standards and their alignment are shown in the tables in the appendix. Modeling and advanced standards are not treated in separate chapters but are instead discussed within the context of the relevant subject category.

Alignment

Following the tables, we will briefly summarize the alignment with the standards. If a standard is partially aligned—that is, the document considered it aligned, but with a qualification, or it only considered a portion of the standard to be aligned—we will list which segments do and do not align in the alignment section directly after the table.

Discussion

Here we discuss why and how the standards align and don't align. For those that don't align well with the test itself, we discuss how the standards might align to the skills and techniques used during the act of preparation. Furthermore, we discuss ways to incorporate material from the test into tasks beyond the scope of the test. Note the original alignment document often does not give detailed explanations about why a standard does or does not align with the test. Any discussion of such here is our own judgment based on our extensive experience and knowledge of the SAT.

Summary

We give a summary of the previous discussion. This summary may also list suggestions for how to use test material in ways beyond the test's scope in order to meet a standard that otherwise does not align.

Sample SAT Questions

When a domain contains standards that align with the test, we provide some sample questions that demonstrate the concepts from the standards in that domain. All sample questions come from the four sample tests available for download on the College Board's website, which are the same four tests available in their book, *The Official SAT Study Guide*. All problems are listed with a three-number code, defined as

[test number].[section number].[question number]

Thus, "1.3.15" is test 1, section 3, question 15. Only sections 3 and 4 are Math sections. Section 3 contains 20 questions and section 4 contains 38 questions. Note also that questions are numbered by difficulty, so questions toward the end of a section are generally harder than are those at the beginning.

Alignment Beyond the Test

The amount of alignment relevant to you depends on your goal for your class. If you are simply teaching test preparation for the sake of doing well on the test, you should be concerned with the skills that are required to answer the questions and the skills that will be needed during the test prep classroom activities. These skills alone account for the majority of the standards. This is an important point; you can be confident that *test prep is fully compatible with the Common Core* and helps students acquire and refine real skills they will need and use in college or careers.

Remember also that some standards will align poorly. The test will not perfectly satisfy all the Common Core requirements. Test preparation should not replace your usual math classes, but it can *supplement* them. Test prep gives you more math drills and exercises, all of it fully compatible with the standards. Not every standard aligns with the test, but every question on the test aligns with the standards.

Missing Skills

Since the SAT is a high school test, we're focusing on the high school standards in this discussion. But there are also some concepts on the test that students should have learned *before* high school. For example, computing a simple percentage is a skill that shows up on the SAT in all sorts of questions, but percentages aren't mentioned in any of the standards here. However, computing percentages does show in standards for sixth and seventh grade. So if you encounter an SAT question that doesn't seem to have a matching standard, it may align to an earlier standard.

Standards for Mathematical Practice [MP]

1. **Make sense of problems and persevere in solving them.**
2. **Reason abstractly and quantitatively.**
3. **Construct viable arguments and critique the reasoning of others.**
4. **Model with mathematics.**
5. **Use appropriate tools strategically.**
6. **Attend to precision.**
7. **Look for and make use of structure.**
8. **Look for and express regularity in repeated reasoning.**

Alignment

The SAT is fully aligned with all Standards for Mathematical Practice.

Discussion

The Standards for Mathematical Practice are well aligned to the SAT. These skills are necessary for questions throughout the test material. In fact, it's skills such as these that can make the test so difficult for so many students, and they were at the forefront of our thoughts when we designed our test preparation system.

Many of these standards are directly or indirectly addressed in the "Math Techniques" chapter of *The Book of Knowledge*. The "General Strategies" section outlines strategies to help students piece together and understand confusing problems. It has strategies to help students read carefully and eliminate careless mistakes. It has strategies that help students switch between abstract reasoning and quantitative reasoning. It has strategies with concrete structures that are repeatable across many different question types. In short, it specifically addresses almost all of the standards listed here.

Let's go through all of them one by one.

1. Make sense of problems and persevere in solving them.

2. Reason abstractly and quantitatively.

Standards 1 and 2 are obviously true on their face for all SAT Math problems. This is what it means to solve math problems.

3. Construct viable arguments and critique the reasoning of others.

The College Board considered this to be aligned with the old test, but it's a bit of a stretch. The second part, about critiquing others, doesn't really apply to SAT Math problems. Reading problems, sure, but not math so much. The first part, "construct viable arguments", is applicable, albeit only somewhat. Students won't be asked to literally construct arguments. There are no proofs, and you don't have to justify your answer. However, they will have to reason and form conclusions, reason inductively about data, and determine multistep procedures for solving a problem.

4. Model with mathematics.

There is a significant amount of mathematical modeling on the SAT, including but not limited to creating functions based on contexts in plain English and making relationships between functions, tables, and graphs.

5. Use appropriate tools strategically.

There aren't many tools allowed on the SAT, but they do get calculators (sometimes). While calculator use is forbidden on one section, questions where calculators are permitted place no restrictions on how students use calculators. They may use them as much or as little as they deem appropriate.

6. Attend to precision.

Yes, yes, yes, absolutely. Carelessness is a huge problem for some students. They have a solid foundation of mathematical knowledge but constantly make small mistakes and get questions wrong, often because of timing concerns. Questions are written with distractor choices that are intentionally tempting and trip up students who are not careful.

7. Look for and make use of structure.

SAT questions often test structure, explicitly and implicitly. Algebra questions in particular will ask students to rewrite the structure of equations in order to highlight a particular quantity or a feature of a graph. Indirectly, more complicated questions will require students to see structural similarities or connections that aren't immediately apparent, such as treating multiple variables together as single units, or rearranging given equations to find familiar identities.

8. Look for and express regularity in repeated reasoning.

SAT problems often ask students to notice patterns. For example, questions may present a table of values and ask students to determine if they represent linear or exponential growth, or to write an equation describing the relationship between the variables.

Summary

The SAT is strongly aligned with the Standards for Mathematical Practice. Furthermore, A-List's math techniques encourage the development and refinement of these very skills.

Number and Quantity [N]

The Real Number System [N-RN]

- *Extend the properties of exponents to rational exponents. (2)*
- *Use properties of rational and irrational numbers. (1)*

Quantities [N-Q]

- *Reason quantitatively and use units to solve problems. (3)*

The Complex Number System [N-CN]

- *Perform arithmetic operations with complex numbers. (3)*
- *Represent complex numbers and their operations on the complex plane. (3)*
- *Use complex numbers in polynomial identities and equations. (3)*

Vector and Matrix Quantities [N-VM]

- *Represent and model with vector quantities. (3)*
- *Perform operations on vectors. (2, 5 subskills)*
- *Perform operations on matrices and use matrices in applications. (7)*

Alignment

The SAT is aligned with all standards in The Real Number System and Quantities domains. In The Complex Number System, it is aligned only with standards 1, 2, 7, 8, and 9. It does not align with any standards in the Vector and Matrix Quantities domain, except N-VM.3.

Discussion

What Aligns

The SAT contains a number of questions that explicitly deal with the real number system and quantities.

The test will contain problems involving properties of exponents, both integer and rational exponents, and will require students to know and employ rules for their manipulation. The test makes use of irrational numbers, such as certain square roots or π.

The test also will contain problems that require quantitative reasoning and use of appropriate units. Certainly, it has always been common for problems to employ units, but the redesigned SAT now contains more problems in which students will have to convert units (often with conversion ratios provided), sometimes in several steps (such as converting miles per hour to kilometers per second). Students will also have to interpret graphs that contain disparate or nonlinear scales on their axes.

One of the changes to the new redesigned test was the addition of problems involving imaginary and complex numbers. Such questions usually are fairly straightforward. Students will have to understand the definition $i^2 = -1$ and the value of i taken to different powers (standard N-CN.1). Students will have to manipulate polynomials involving complex numbers in all the same ways they do algebraic polynomials (standard N-CN.2), as well as algebraic identities such as the difference of squares (standard N-CN.8).

Standard N-CN.7, "Solve quadratic equations with real coefficients that have complex solutions," is an interesting case. Finding complex solutions for quadratics is not specifically listed among the skills in the new SAT's specifications document, and no practice problems have asked students to do so. However, there are problems asking to find exact complex solutions to quadratic equations among the practice problems at Khan Academy. These problems were created with direct help and approval of the College Board, so its presence there indicates that it's a concept that may well appear on the real SAT in the future. Given that the redesigned test has increased frequency and complexity of quadratic equations and the addition of complex numbers, it seems very likely that will be the case.

Additionally, the alignment document for the old SAT listed N-CN.9 as aligned—the only standard in the domain to be so. This standard refers to the Fundamental Theorem of Algebra, which is relevant to some SAT problems (even if students don't have to know specifically what it is). Since it was considered aligned for the old test, and the algebra content of the test has only increased, we are keeping it aligned now.

It should be noted that complex number questions are very infrequent. These questions fall into the "Additional Topics in Math" category, of which there are only six questions per test. Of those six, most will be geometry questions; only one, if any, will likely involve complex numbers.

What Doesn't Align

All complex number standards besides those mentioned above are outside the scope of the SAT. These include concepts that involve graphing complex numbers in the complex plane and finding moduli.

Matrices and vectors are outside the scope of the SAT. However, the alignment document for the old SAT did consider N-VM.3 to be aligned. This standard refers to "problems involving velocity and other quantities that can be represented by vectors", which certainly appear on the SAT, albeit not in vector form.

While none of the released practice tests contained anything resembling vector format, there have been a few questions in the Khan Academy problem sets that use vectors. These are simple questions that are little more than right triangle questions requiring the Pythagorean Theorem. We are being conservative here in considering the vector standards unaligned. We concede that they might appear more explicitly in the future, but even if they do, it would be rarely and would likely not require much knowledge of vector operations.

It should be noted that all of the standards in this domain are advanced standards. All of the College and Career Ready standards are aligned.

Summary

The SAT will contain problems involving exponents with both integer and rational values and will make use of rational and irrational numbers. The test will require quantitative reasoning and test conversion of units. It will contain problems involving complex numbers.

The SAT will not test knowledge of matrices or vectors, but it will use concepts that can be represented as vectors.

Sample SAT Questions

The Real Number System: 1.3.14, 2.3.7, 3.3.3

Quantities: 1.4.6, 2.4.11, 3.4.14, 4.4.33

Complex Numbers: 1.3.2, 2.3.11, 4.3.14

Algebra [A]

Seeing Structure in Expressions [A-SSE]

- *Interpret the structure of expressions. (2, 2 subskills)*
- *Write expressions in equivalent forms to solve problems. (2, 3 subskills)*

Arithmetic with Polynomials and Rational Expressions [A-APR]

- *Perform arithmetic operations on polynomials. (1)*
- *Understand the relationship between zeros and factors of polynomials. (2)*
- *Use polynomial identities to solve problems. (2)*
- *Rewrite rational expressions. (2)*

Creating Equations [A-CED]

- *Create equations that describe numbers or relationships. (4)*

Reasoning with Equations and Inequalities [A-REI]

- *Understand solving equations as a process of reasoning and explain the reasoning. (2)*
- *Solve equations and inequalities in one variable. (2, 2 subskills)*
- *Solve systems of equations. (5)*
- *Represent and solve equations and inequalities graphically. (3)*

Alignment

The SAT is aligned with all standards in the Algebra conceptual category except A-REI.8 and A-REI.9.

Discussion

What Aligns

The test includes a wide variety of algebra questions in substantial quantities. Over half of the test will deal directly with concepts from algebra between the two major content categories: Heart of Algebra (33%) and Passport to Advanced Math (28%). These concepts include solving simple one-variable equations, solving multivariable systems of equations, manipulating and understanding polynomials, solving inequalities, and creating and solving equations based on settings expressed in words. This last concept is particularly common and directly aligns with the kind of "modeling" that is so important to the Common Core math standards. All of these topics are discussed in the respective algebra sections of *The Book of Knowledge*.

One of the major features of the redesign was an increase in the amount of algebra on the test; the old test only had about 30%–40% algebra questions. This not only means an increase in the frequency of algebra questions, but also an increase in the number of skills involved and the potential difficulty of the questions. For example, the test now contains questions in which students must divide polynomials leaving an algebraic expression as a remainder, which would never have appeared on the old test.

In addition, there are more questions that explicitly ask students to understand the parts of equations and why they are set up a certain way. For example, the old test would often present word problems that describe relationships and then ask students to choose an equation that properly models that relationship. The new test will ask such questions, too. But you may also see questions in which the relationship's equation is given to you and you're asked something like "what does the number 12 in the equation represent?"

These changes all reflect a close alignment to Common Core algebra. In particular, the CCSS emphasize an understanding of the structure of equations and a fluency in manipulating them, both skills that are highlighted more on the redesigned test.

However, despite these updates to the test, the total algebra alignment to the Common Core has not changed from the old test. This is partly because the old test's alignment document was somewhat liberal in what it considered aligned; rather than being aligned to more standards, the redesigned test is aligned to the same standards, but more strongly.

For example, A-SSE.3c, the topic of writing expressions in equivalent forms, states: "Use the properties of exponents to transform expressions for exponential functions. *For example, the expression* 1.15^t *can be rewritten as* $(1.15^{1/2})^{12t} \approx 1.012^{12t}$ *to reveal the approximate equivalent monthly interest rate if the annual rate is 15%.*" The original alignment document described this as aligned, even though exponential functions rarely appeared on the old test, and students were never asked to manipulate them in the way described in the example. However, the redesigned test frequently asks about exponential questions (about two questions per test on average). Furthermore, such questions often ask students to understand the structure of exponential functions in order to be able to manipulate them in the way described in the example. So, from the College Board's perspective, this standard was aligned before and is aligned now, so there's no change in the number of aligned standards. But in practice, it was very weakly aligned before and is strongly aligned now.

What Doesn't Align

The test is partially aligned to A-SSE.4 (the alignment document includes the note, "The SAT aligns to use of the formula to solve problems."), A-REI.5 (with the note, "The SAT aligns to use of the embedded skill to solve problems."), and A-REI.11 (with the note, "Students are expected to be able to identify points of intersection as the solution on the SAT. They may use graphing calculators to determine the answers."). In each of these cases, the standard asks students to "derive" or "prove" a concept. The College Board's alignment document states that the SAT aligns for the *use* of the concepts mentioned in the standards, but the test does not require proofs or derivations. On the SAT, only the answer matters, not the method.

This indifference to method is important to note, since the heavy emphasis on algebra can be problematic for students who struggle with algebra or with abstract reasoning. As we mentioned, there is often more than one way to get the right answer. A-List's math techniques, as outlined in *The Book of Knowledge,* give methods of solving problems specifically geared toward students who struggle with algebra questions, through not understanding the concepts, working too slowly, or making careless mistakes. All of our techniques are designed to work with concrete values instead of abstract variables. This is not just a trick or a shortcut; it's an effective a way for students to develop an understanding of why algebra is true and how the abstraction of algebra is a generalization of all concrete solutions.

The SAT is not aligned with A-REI.8 or A-REI.9. Both those standards refer to matrices, which do not explicitly appear on the SAT. For standard 8, the alignment included the note, "The SAT does not require that students represent systems in matrix form exclusively." The SAT often contains questions asking students to solve systems of equations, which *can* be solved using matrices, but matrix format is neither mentioned in the question nor required to solve it.

In fact, this point applies to many of the standards here; they discuss algebraic concepts that *can* be used on either test, but aren't *necessary* to solve the questions. Given that you *could* use this method, it's somewhat strange that the alignment document doesn't consider this to be aligned. There are other standards that have a much weaker claim to relevance that the document did consider aligned (see Statistics and Probability, for example). The old test did feature systems of equations, but the redesigned test features them more frequently, so there's an argument for calling these aligned.

Summary

The SAT is strongly aligned to all the algebra domains. As usual, the test does not demand that students use particular methods for solving problems, but nearly all the concepts listed in these standards are available to them.

The test does not require students to derive or prove algebraic truths, but will require students to use those concepts. It also does not require knowledge of matrices.

Sample SAT Questions

Structure of Expressions: 1.4.30, 2.3.3, 3.3.12, 4.4.14

Polynomials and Rational Expressions: 1.4.29, 2.3.15, 3.4.6, 4.4.12

Creating Equations: 1.3.3, 1.4.9, 2.3.14, 3.3.4

Reasoning With Equations and Inequalities: 1.3.11, 2.4.29, 3.4.30, 4.3.15

Functions [F]

Interpreting Functions [F-IF]

- *Understand the concept of a function and use function notation. (3)*
- *Interpret functions that arise in applications in terms of the context. (3)*
- *Analyze functions using different representations. (3, 7 subskills)*

Building Functions [F-BF]

- *Build a function that models a relationship between two quantities. (2, 3 subskills)*
- *Build new functions from existing functions. (3, 4 subskills)*

Linear, Quadratic, and Exponential Models [F-LE]

- *Construct and compare linear, quadratic, and exponential models and solve problems. (4, 3 subskills)*
- *Interpret expressions for functions in terms of the situation they model. (1)*

Trigonometric Functions [F-TF]

- *Extend the domain of trigonometric functions using the unit circle. (4)*
- *Model periodic phenomena with trigonometric functions. (3)*
- *Prove and apply trigonometric identities. (2)*

Alignment

The SAT is aligned with all the standards in the Interpreting Functions and Building Functions domains. It is aligned with almost all of the standards in the Linear, Quadratic, and Exponential Models domain, with the exception of F-LE.4, to which it is not aligned. In the Trigonometric Functions domain, it is aligned only with standards 1, 2, 3, and 4.

Discussion

What Aligns

The SAT has a substantial number of questions that explicitly test knowledge of functions and function notation. These questions come in several different forms. Some will provide equations in *f*(*x*) format, asking students to evaluate, manipulate, or transform those equations. Some will provide similar equations in the context of an actual setting. Others still will ask students to produce, interpret, or transform graphs of functions. Function notation generally appears on Passport to Advanced Math questions, which make up 28% of the test (though not all questions in that subtype will use function notation). Most of the time, these questions are little different from algebra questions, but students will certainly be required to understand how functions work.

Furthermore, it should be noted that A-List's math techniques like Plug In and Backsolve work just as well for function problems as they do for algebra problems. These techniques can be particularly useful for intricate questions for which students easily get lost, such as those involving inverse functions, nested functions, or function transformations.

Just as with algebra, the redesigned test doesn't align with more standards so much as it strengthens the existing alignment. As we've seen, the official alignment documents are occasionally somewhat liberal about considering standards to be aligned. For example, the SAT alignment document stated that the old test aligned with standard F-IF.7, which requires students to graph a number of different types of functions. But some of the skills listed there—particularly cube root, rational, and exponential functions—were beyond the scope of the skills on the old test. Most function graphs on the old SAT were simple lines or parabolas. However, the new SAT does include a wider range of functions and asks students to understand their graphs. So the new test is aligned with the same number of standards, but is aligned more strongly.

One place where the new test is aligned with more standards is trigonometry, which is in much the same boat as complex numbers mentioned in the Number and Quantity section. It's one of the new concepts that have been added to the redesigned test, but, like complex numbers, trig questions occur infrequently. They belong to the Additional Topics in Math content category, of which six questions will appear, most of which will be geometry questions. You can expect to see about one, maybe two, trigonometry questions on any given test.

Because it's a new addition to the test, all trigonometry standards were marked as unaligned in the College Board's study for the old SAT. Therefore, we have to do some guesswork to determine which of the standards will align and which won't. Questions will definitely ask students to know the definitions of the main trigonometric functions and their relationship to angles and triangles. They will be expected to know how to convert radians to degrees and vice versa and the ratios of special right triangles. The unit circle has appeared only indirectly in the official practice tests, but it has appeared frequently on Khan Academy materials, so it's clearly an important concept. Those concepts cover the four standards in the first cluster of the Trigonometric Functions domain. The rest of the standards are outside the scope of the SAT.

What Doesn't Align

Standard F-LE.4 does not align because the SAT does not test knowledge of logarithms.

Based on the information available, the standards in the second two clusters of the Trigonometric Functions domain do not align well with the test. The released material does not include any questions involving graphing trigonometric functions on the *xy*-plane (besides the unit circle), nor any questions addressing the periodicity of the wave graphs. The periodicity of the functions has shown up only indirectly, inasmuch as it's necessary to understand the unit circle. Nor have there been questions involving inverse functions or algebraic identities such as $\sin^2(\theta) + \cos^2(\theta) = 1$.

Standard F-TF.9 is a borderline case. Trig questions on the SAT have not required addition formulas such as $\sin(a + b) = \sin(a)\cos(b) + \sin(b)\cos(a)$. However, there have been a few questions requiring students to know $\sin \theta = \cos (90 - \theta)$. We marked this as unaligned, since trig questions as a whole have not required much knowledge of equations beyond the function definitions. But for this—and for the other trigonometry standards mentioned above—we would not be surprised if we start to see some additional trigonometric concepts once we get a larger data set of real test questions.

Summary

The SAT requires students to be able to understand function notation and to build, manipulate, evaluate, and graph functions of many different types. Students will be expected to know and use the main trigonometric functions, though they will generally not have to graph them, nor will they need more complex identities.

Sample SAT Questions

Interpreting Functions: 1.3.10, 3.4.16, 4.3.2, 4.4.28

Building Functions: 1.3.3, 2.4.10, 4.3.4, 4.4.25

Linear, Quadratic, and Exponential Models: 1.4.37, 3.4.4, 3.4.21, 4.4.13, 4.4.20

Trigonometric Functions: 1.3.19, 2.3.19, 3.4.23, 4.3.17

Geometry [G]

Congruence [G-CO]

- *Experiment with transformations in the plane. (5)*
- *Understand congruence in terms of rigid motions. (3)*
- *Prove geometric theorems. (3)*
- *Make geometric constructions. (2)*

Similarity, Right Triangles, and Trigonometry [G-SRT]

- *Understand similarity in terms of similarity transformations. (3, 2 subskills)*
- *Prove theorems involving similarity. (2)*
- *Define trigonometric ratios and solve problems involving right triangles. (3)*
- *Apply trigonometry to general triangles. (3)*

Circles [G-C]

- *Understand and apply theorems about circles. (4)*
- *Find arc lengths and areas of sectors of circles. (1)*

Expressing Geometric Properties with Equations [G-GPE]

- *Translate between the geometric description and the equation for a conic section. (3)*
- *Use coordinates to prove simple geometric theorems algebraically. (4)*

Geometric Measurement and Dimension [G-GMD]

- *Explain volume formulas and use them to solve problems. (3)*
- *Visualize relationships between two-dimensional and three-dimensional objects. (1)*

Modeling with Geometry [G-MG]

- *Apply geometric concepts in modeling situations. (3)*

Alignment

The SAT is aligned with all standards of the Geometry conceptual category except for standards G.SRT.9, G-SRT.10, or G-SRT.11.

Discussion

What Aligns

First and foremost, almost all of the standards align with the SAT. The SAT features geometry problems using a wide range of topics including but not limited to angles, circles, triangles, polygons, area, volume, and geometry on the coordinate plane. However, simply saying that 93% of the geometry standards are aligned may paint an inaccurate picture, because geometry is a very small part of the test.

Geometry is in a tough spot on the redesigned SAT. The old test had roughly equal amounts of algebra and geometry. The College Board wanted to focus more on algebra, which meant that much of the geometry was cut. This wasn't done cavalierly; it was in response to data from postsecondary instructors about which concepts are most applicable to their coursework. Since the SAT is a college admissions test, it is important to the College Board for the SAT to test those skills most relevant to colleges. (The test specifications document about the redesigned test discusses this further in the "Evidentiary Foundation for the Redesigned Test" section.)

But geometry is clearly a large part of the Common Core—about a quarter of the standards fall into this category. While in other areas we've seen efforts to make the SAT more closely aligned with the Common Core, here is an area where the College Board has consciously made the test diverge from the CCSS because it prioritizes the test's identity as a college admissions test.

Where this leaves us is that the distribution of concepts in the CCSS does not match their distribution on the SAT. Of the 58 math questions, six questions will be in the content category "Additional Topics in Math" (the only category for which students do not get a subscore). Of those six, about five questions will be geometry questions, or about 9% of the test. That's a far cry from the old test, where geometry questions made up 25%–30% of the test.

However, no particular concepts were *cut* from the test. All the same geometric concepts from the old test may still appear on the new test—so all the standards still align—they just appear in smaller numbers. Some of the geometry problems that do appear can be quite complex and require a synthesis of disparate rules. In fact, this category is actually slightly more aligned with the CCSS now, since the redesigned test has added trigonometry, so standard G-SRT.7, previously unaligned, is now aligned.

As a rule, figures on the SAT are always drawn to scale unless the figure explicitly states otherwise. This is rather important with regard to one of our general math strategies, Guesstimate. It exploits the fact that figures are drawn to scale on the SAT and asks students to use their skill at observing and visually evaluating measurements directly to approximate the answer in increments of increasing accuracy. That is: look at the picture and eliminate choices that are obviously too big or too small.

What Doesn't Align

There are a number of standards that are partly aligned, in that the College Board's alignment document included qualifying statements. The test is partially aligned to G-CO.9, G-CO.10, G-CO.11, G-SRT.4, and G-C.1, all with the same note: "The SAT aligns to the use of the concepts within the proof to solve problems." It is partially aligned to G-C.3 with the note, "The SAT aligns to understanding of the properties of angles of the quadrilateral inscribed in a circle." It is partially aligned to G-C.4 with the note, "The SAT aligns to understanding of the properties of a line outside a circle tangent to the circle." It is partially aligned to G-GMD.1 and G-GMD.2 with the same note for both, "The SAT aligns to the concepts within the standard but does not require construction of an informal argument."

The standards that are listed as partially aligned with the SAT all deal with the same issue. The geometry standards frequently ask students to construct proofs of certain theorems or to derive particular formulas using other theorems or rules. This kind of work is grounded in the classical Euclidean tradition of geometry and is often the focus of geometry classes in school. The SAT is a results-based test and is only concerned with the solution to a particular problem. Thus, students will never be asked to explicitly prove or derive any geometrical proof. That's not to say, however, that these skills are unimportant for the test. On the contrary, the SAT, rather than asking students to know a lot of things superficially, asks students to know a *few* things *really well*. Indeed, this is why many students struggle with the test. Therefore, being able to prove theorems rather than simply memorize them can give students a much deeper understanding of the underlying truths and will make employing the theorems when necessary that much easier.

The SAT is unaligned with standards 9, 10, and 11 in the Similarity, Right Triangles, and Trigonometry domain. As we've mentioned, the redesigned test does now include trigonometry (which is why G-SRT.7 is now aligned), but there has been no indication that students will need the formulas listed in these three standards, such as the Law of Sines or the Law of Cosines.

Summary

The SAT contains geometry questions in all the domains listed above, but not in great numbers. The test will not require students to prove or derive any formulas or theorems, but the ability to do so can give students a deeper understanding of the necessary geometric principles.

Sample SAT Questions

Congruence: 1.4.3, 3.3.11, 3.3.18

Similarity, Right Triangles, and Trigonometry: 1.3.17, 2.3.18, 3.4.23, 4.3.17

Circles: 2.4.36, 3.4.34, 4.4.24, 4.4.36

Expressing Geometric Properties With Equations: 1.4.24, 2.4.24, 2.4.28, 4.3.8

Geometric Measurement and Dimension: 1.4.35, 3.4.25, 4.4.18

Modeling With Geometry: 1.3.17, 4.3.16, 4.4.18

Statistics and Probability [S]

Interpreting Categorical and Quantitative Data [S-ID]

- *Summarize, represent, and interpret data on a single count or measurement variable. (4)*
- *Summarize, represent, and interpret data on two categorical and quantitative variables. (2, 3 subskills)*
- *Interpret linear models. (3)*

Making Inferences and Justifying Conclusions [S-IC]

- *Understand and evaluate random processes underlying statistical experiments. (2)*
- *Make inferences and justify conclusions from sample surveys, experiments, and observational studies. (4)*

Conditional Probability and the Rules of Probability [S-CP]

- *Understand independence and conditional probability and use them to interpret data. (5)*
- *Use the rules of probability to compute probabilities of compound events in a uniform probability model. (4)*

Using Probability to Make Decisions [S-MD]

- *Calculate expected values and use them to solve problems. (4)*
- *Use probability to evaluate outcomes of decisions. (3, 2 subskills)*

Alignment

The SAT is aligned with all the standards in all four domains of the Statistics and Probability conceptual category with the exception of S-ID.8 and S-MD.2.

Discussion

What Aligns

The redesigned SAT features an increase in the number of questions dealing with statistics. On the new test, 29% of the questions belong to Problem Solving and Data Analysis. Of those, about a third are questions about proportions, ratios, and rates, while the rest relate to statistics and data analysis, which means about 15%–20% of the test will feature statistical concepts. On the old test, only 10%–15% of the questions on the SAT belong to a category called "Data Analysis, Statistics, and Probability".

The new test also draws from a larger set of concepts in this realm. The old SAT's statistical questions dealt with only a few basic concepts: averages, basic probability, combinatorics, and reading a table or chart. The new test features questions on more advanced concepts, including but not limited to:

- Analyzing the center and spread of data sets in multiple ways.
- Interpreting histograms, frequency tables, and scatterplots with trend lines.
- Making estimates and projections based on sample data.
- Drawing conclusions and inferences from data or evaluating the appropriateness of such conclusions.
- Calculating probabilities in data sets with multiple variables.

These concepts broadly match well to the domains for the CCSS in this category.

It is surprising that the College Board's alignment considered the old SAT 94% aligned with the CCSS in this category. Maybe not "surprising" so much as "ridiculous". The old test had a very rudimentary statistical component, and these standards list specific, sometimes advanced, skills that had little to do with the questions on the test. The majority of such questions involved little more than *reading* a graph, or *computing* a probability, with very little actual *analysis* of the data demanded by, say, standard S-ID.9, "Distinguish between correlation and causation." Just like we've seen on the other categories, the redesigned SAT doesn't align with *more* standards than the old test did (or claimed to), but it does align more *deeply*. A standard like, for example, S-ID.3, "Interpret differences in shape, center, and spread in the context of the data sets, accounting for possible effects of extreme data points (outliers)" is something that was only indirectly relevant to the old test but could be explicitly asked on the redesigned test.

What Doesn't Align

Standards S-ID.8 and S-MD.2 were unaligned on the old test and remain so on the new test; the new test will not ask for correlation coefficients, nor ask to calculate expected values of random variables.

While some standards that were unjustifiably considered aligned on the old test are now solidly aligned, others are still questionable. Most of the Using Probability to Make Decisions domain is beyond the scope of the test. We're listing them as aligned here partly because they were listed that way before and partly because they involve reasoning skills that indirectly relate to the test. For example, all standards in the cluster "Use probability to evaluate outcomes of decisions" give specific scenarios that are unlike anything on the SAT, such as S-MD.6, "Use probabilities to make fair decisions (e.g., drawing by lots, using a random number generator)." Nothing on the SAT involves making decisions like this. However, there are questions asking to calculate probabilities that could be extended to decision making, so the underlying mathematical skill aligns, even if the literal scenarios don't.

Summary

The SAT aligns very strongly with the Statistics and Probability category and contains a large number of problems dealing with data visualization, statistical analysis, and probability.

Sample SAT Questions

Interpreting Categorical and Quantitative Data: 1.4.14, 2.4.18, 3.4.20, 4.4.21

Making Inferences and Justifying Conclusions: 2.4.13, 2.4.20, 3.4.15, 4.4.4

Conditional Probability and the Rules of Probability: 1.4.21, 2.4.16, 3.4.29, 4.4.9

3

SAT Math in the Classroom

Overview

We know that SAT Math correlates with Common Core math. What does that mean for you the teacher? How can you get your math classes ready for the SAT?

Some of you may have a full SAT prep course set up as its own class in your school, either as an elective or after school. If you have time to dedicate exclusively to test prep, it's best to get some materials specifically designed for such a course. A-List not only has our own materials for such programs, we also offer professional development and assistance in setting up your courses.

Full prep classes should be reserved for junior year. You want to make sure (a) that they've been exposed to all the math they'll need for the test, and (b) that the classes are given close to the time when students will take the real test. If a prep class ends in June but the students don't take the real test until October, they will surely forget everything you've done in the meantime.

Not everyone has the time and resources available for dedicated prep courses. But that's okay! This section will show you how you can weave SAT math material into your normal math classes in ways that can enhance the material you're already using.

When you're preparing students for the SAT, there are three areas of the test you should consider: *Content*, *Techniques*, and *Timing*. We'll go through a number of ways to expose your students to practice problems, depending on which of these three areas you want to focus on.

Content

The first thing you can do is *don't change a thing*. Unless you're teaching calculus or other concepts beyond algebra 2, everything you're already doing is going to be relevant for the SAT to some degree. Not all concepts will be equally present, but that's inevitable with the constraints of the test. It's tough to squeeze three years of math into 58 questions. But just about everything you do will contribute somehow.

The easiest thing to do outside of a prep course is to simply weave sample SAT material into your courses. You can do so with material reflecting the work you're currently doing anyway, with work you've done recently this year, or surprise them with work from previous years.

How much you do is up to you. If you're working on content that is underrepresented on the SAT, then you'll probably want to scale back the amount of material. Geometry, for example, is less than 10% of the test. When you're teaching a geometry course, the stuff you're working on will show up, just not very often. So an occasional problem or two is useful, but not every day. If you're doing algebra of linear functions, that stuff is all over the test.

Of course, you don't have to give problems that are directly relevant to today's lesson. It's great to call back to math concepts they've already learned (and forgotten). Paradoxically, sometimes students struggle more with the *easier* math because it's been too long since they've had to think about it. You've just spent the last week talking about parabolas, so they're fresh in your students' minds, but it's been years since anyone has asked them to add fractions without a calculator. As students approach the end of high school, they remember less and less of the beginning of high school. We've seen so many students miss questions that require little more than sixth-grade math.

The other side of this is that the test may include some content that isn't in your curriculum at all. This may be true, for example, if your school doesn't have a dedicated statistics course, or if some more advanced algebra or trigonometry concepts don't show up until senior year. If that's the case, you may consider bringing them up specifically in the context of the SAT. Statistics, for example, show up on the test in a variety of contexts, but never in great detail. You may see a scatterplot with a line of best fit, but you'll simply have to know what that means. You won't have to draw the line or calculate *R*-squared.

Techniques

Content is only one part of the test. There's also a question of *techniques*, figuring out the best way of doing problems. A-List uses two main math techniques: **Plug In** and **Backsolve**. They both have the same goal: to turn algebra problems into arithmetic problems.

- Plug In exploits the fact that an equation is true for all inputs (usually). If the answer choices contain algebraic expressions, then the right answer will always be right, no matter what specific values you choose. So pick a number for your variable, put it through the problem, and get another number as your answer. Put your made-up number in the choices and see which gives you the same output.
- Backsolve works for questions where there only is one possible value. If there are numbers in the answer choices, pick a choice and make that number the answer to the question. Put the number through all the information you have and make sure everything matches. If it does, that's your answer. If it doesn't, pick a different choice.

Our textbook, *The Book of Knowledge*, discusses the math techniques in greater detail. At heart, they are rather straightforward and don't take long to understand. But they do require a lot of practice to do well. This is true of literally all new skills. Just because you can read sheet music doesn't mean you can play the piano. You have to practice in order to execute.

You can introduce these techniques to your students in just a few minutes of class time. Once you do, you can bring them back up constantly. Any time you do a practice problem, ask if it can be done with a technique. Sometimes the answer is no, but overall about 40% of SAT questions can be done with Plug In or Backsolve. It takes practice to get comfortable using methods beyond what you're used to, but these methods are crucial to score improvement. If your students don't change their methods, they won't change their scores.

Timing

The last element of the test is putting the content and techniques together under a time limit. So many students have problems that stem from time management. The only way to work on these issues is with some timed practice.

Timed Practice

Full Tests

The best way to expose students to the test is to give real, proctored practice tests. Any exposure to practice problems is valuable for students, but full-length proctored tests are best for a number of reasons:

- They give students exposure to the test format and conditions.
- Practice tests given under real-test conditions produce scores that most accurately predict real scores.
- They allow students to practice timing and time-sensitive test-taking techniques.

Timing can be a huge issue for students, and even if they know all the math perfectly, it can be difficult for them to acclimate to the high-pressure 3-hour test format. Anyone with experience teaching the test has seen students who do much better on problems done leisurely without time constraints than on a timed section—to say nothing of a timed section after they've already sat through more than an hour and a half of reading and writing. The most common problems that arise include:

- Students move too slowly and run out of time before completing as many questions as they want.
- Students move too quickly and make careless mistakes on questions they know how to do.
- Pure mental fatigue of working for 3 hours negatively affects students' performance.

This is often what holds back students who normally do well in school but don't score as highly as they expect on the SAT. It's not that they don't know enough math, it's that they're missing questions they already know how to do. They know most of the math, but they struggle to apply it efficiently.

So what do we do about timing problems? It's something you can discuss explicitly with your students, particularly as you get closer to the real test. A-List's *The Book of Knowledge* contains several discussions that are specifically geared toward time management. There are two main ways to address time management.

1. **Target numbers**. Unless you're trying to get an 800, you do not actually have to do every question. You'd be surprised how few questions you need in order to get the score you want. If you're trying to get, say, a 600, you only need to get about 40 questions right. If you do *fewer* questions but with *greater accuracy*, it's easier to get your target score. Because the questions are ordered by difficulty, you know you're more likely to get the questions at the beginning. Skip the last questions, spend more time on the first questions, and you'll cut down on careless mistakes. (Just remember to fill random answers for the questions you skip. You don't lose points for wrong answers so you'll pick up some points from random guessing.)
2. **Math techniques**. As we mentioned earlier, the math techniques, Plug In and Backsolve, are ways to turn algebra problems into arithmetic problems. That often means making questions easier and making them go faster. More importantly, they help cut down on careless mistakes, so it makes you more accurate, and accuracy is our goal for using target numbers.

Math techniques can be practiced on any problem set, but in order to practice target numbers, you have to give timed sections.

The PSAT

It can be difficult to manage the logistics administering a practice test—few schools have an extra 3 hours lying around available for use. Many schools already give a practice test in the form of the PSAT. The PSAT (or "PSAT/NMSQT" for "National Merit Scholarship Qualifying Test") is an official test administered by the College Board. It does not count towards college admissions, but it is a good predictor of a student's SAT score. It's just like the SAT with two main differences: there's no essay, and each section is slightly shorter. Schools and students will receive detailed reports about students' performance. As the name implies, students with the highest scores can qualify as National Merit Scholarship semifinalists.

Most students will take it their junior year, but the College Board also offers two options for younger students: the PSAT 10 to sophomores and the PSAT 8/9 to eighth and ninth graders. The PSAT 10 is identical to the junior-year PSAT/NMSQT, but the PSAT 8/9 is slightly shorter. The content levels are also slightly

Table 3.1 SAT, PSAT, PSAT 8/9 Format

Section	SAT		PSAT/NMSQT and PSAT 10		PSAT 8/9	
1. Reading	52 Q	65 min	47 Q	60 min	42 Q	55 min
2. Writing	44 Q	35 min	44 Q	35 min	40 Q	30 min
3. Math (No calc.)	20 Q	25 min	17 Q	25 min	13 Q	20 min
4. Math (Calc. OK)	38 Q	55 min	31 Q	45 min	25 Q	40 min
5. Essay (optional)		50 min	None		None	
Total	154 Q w/essay	180 min 230 min	139 Q	165 min	120 Q	145 min

Table 3.2 SAT, PSAT, PSAT 8/9 Math Content

Section	SAT		PSAT/NMSQT and PSAT 10		PSAT 8/9	
Heart of Algebra	19 Q	33%	16 Q	33%	16 Q	42%
Problem Solving and Data Analysis	17 Q	29%	16 Q	33%	16 Q	42%
Passport to Advanced Math	16 Q	29%	14 Q	29%	6 Q	16%
Additional Topics in Mathematics	6 Q	10%	2 Q	4%	0 Q	0%

different to adjust for grade levels. It's barely noticeable between PSAT/NMSQT and SAT, but more so between PSAT 8/9 and SAT. You won't see as many "Passport to Advanced Math" questions on PSAT 8/9.

With the redesign, the SAT and PSAT are more similar to each other in format than they once were. The old PSAT was 48% shorter than the old SAT (130 min. vs. 225 min.), while the new PSAT is only 28% shorter than the new SAT (165 min. vs. 230 min.). And that's assuming you take the optional SAT essay. Comparing only the four main sections of the new test, the PSAT is only 8% shorter (165 min. vs. 180 min.). So the PSAT has virtually identical timing to the SAT, and it's a great opportunity to practice timing strategies.

Practice Sections in Class

Even if you are giving a full test-prep course in school, it can be tough to schedule full-length practice tests. As an alternative, you may consider having students take a full section in class.

Students get 25 minutes for section 3 (the no-calculator section) and 55 minutes for section 4 (the calculator-permitted section). Section 4 is likely too long to give in one class, but it shouldn't be difficult to give a 25-minute section within a single class period with time to spare.

Please note, however, that there is a substantive difference in math content between the two sections besides the presence of calculators. Problem Solving and Data Analysis questions only appear in section 4, not section 3, which is almost entirely made up of algebra questions. As a result, section 4 also has a lot more questions placed in real-world settings: about two-thirds of calculator questions are setting questions, while only a quarter of calculator questions are. So you may still want to give students a section 4 in order to expose them to those questions.

You could break the section up over two days by having them *only do the multiple-choice questions*. Remember that only 30 of the 38 questions in section 4 are multiple choice—the last eight questions are non-multiple-choice grid-in questions. That means 79% of the section is multiple choice, which proportionally should take about 43 minutes. That's pretty feasible for one class period for many schools. You can then give the remaining eight grid-ins over 12 minutes on the next day.

If you don't have 43-minute periods in your school, you can split the section up more evenly across two days (say, 30 minutes one day and 25 minutes the next). If you do that, it's probably a good idea to give them the full section for both days and let them split it up how they wish, rather than giving them separate packets of, say, questions 1–19 one day and 20–38 the next. The reason is that the questions are ordered by difficulty—question 1 will be really easy and question 30 will be really hard. Therefore, students won't (and shouldn't) spend their time evenly throughout the section. They'll do the first 10 questions a lot faster than the last 10 questions. Instead, you should simply be sure to give them 55 minutes total across the two days and allow them to divide their attention how they see fit.

(Note that the grid-in section is more of a discrete unit, separate from the rest of the section in this regard. Even though the question numbering is continuous, the two parts are arranged by difficulty separately. That means question number 30 will be a *hard* multiple choice but question 31 will be an *easy* grid-in.)

When Should I Give Tests and Sections?

Practice tests can be given at any point in the year to any class of students, depending on the reason you're giving the test. You can give practice tests to students as young as ninth grade. As we've said, on a most basic level it gives students exposure to the format and a sense of where they are in relation to the national and state averages.

First off, remember that because the test is still new, there are not very many official practice tests in print, so you should be judicious about giving them to students. You don't want to give out four practice SATs to your sophomores so that they have none left to do as juniors.

For younger students, practice tests can be more instructive for administrative purposes than to help the students prep. Ninth graders won't do as well on practice tests as juniors will, and it may not be terribly useful to tell students that they aren't doing well on algebra 2 concepts if they haven't taken algebra 2 yet. But it can give the school valuable data about which areas the class as a whole is strong or weak in. This can be particularly important if your state uses the SAT as its statewide assessment and you want to see if your school is on track.

If you're giving a prep course, you should be sure to give a minimum of two practice tests: one at the beginning and one at the end. That way you can see how students have improved and you can gauge whether your course has been effective. The first test is just used for a baseline to see where students start. You can use PSAT or prior real SAT scores for the baseline. The second test should be towards the end of the course, close to the time when they'll take the real test. It should be given after you've gone over most of the necessary concepts.

For sections or partial tests given in class, there's no need to do a baseline test. If you only do part of a test, you won't be able to get SAT-style scores anyway. And even if you do the whole test, the scores won't be as reliable as they would be for a full-length timed test. You're better off using PSAT scores as a baseline.

For any test that isn't purely a baseline test, you should also spend some time going over the questions in class. Giving the questions is fine practice, but we also want students to improve their performance, too. To do that they need to see what they're doing wrong. It's not simply a matter of going over questions they missed; it's also a matter of making sure they're doing questions in the most effective way. In our experience, most students' score improvement comes not from lectures about math concepts and techniques, but from seeing those concepts and techniques in action when reviewing practice tests.

Since any section will contain questions from an assortment of different content areas, it's best to wait to give them until the students have been exposed to a majority of the concepts. It does little good to have them do problems that you know they'll struggle with at best, stare blankly into the void of space at worst. Once you've given them some grounding in the content, then you can start playing around with it in earnest.

Problem-a-Day

Of course, the simplest thing you can do is to give students a real SAT problem periodically in class. It doesn't have to take up a lot of time. It could just be a single question, once a week, a few times a week, or every day.

An SAT question is a great do-now activity, a nice, quick way to open a class. It doesn't take long; depending on the difficulty, you can have the students do the question, review it, and give multiple ways of solving it in less than 5 minutes. Some teachers prefer to give a question at the end of class as an exit ticket. Either way, it needn't take much time out of your class.

This simple act accomplishes some important goals. By exposing students to the problems on a slow but regular basis, it familiarizes them with the test. On a base level, that makes it less scary for them, but it also habituates them to the sort of math they'll see. The test certainly repeats problems types over and over, so the more they see real problems, the more comfortable they'll be at solving them. It also reinforces the idea that SAT math *is not different* from regular math. It's not some special, elusive Fancy Math. It's just plain ol' math, like they do every day.

This sort of activity can be done with students of any age. It's a great thing to start on younger grades. If they see an SAT problem every day for three years, by the time they have to take the real test it will seem practically banal. Furthermore, you can demonstrate to them that the math they're doing in ninth grade is going to be important later on. If they can see that they'll need their ninth-grade math when they get to the SAT in eleventh grade, they're more likely to make an effort to remember it.

Which Problems Should I Choose?

Ideally, you can choose problems that use the same concepts you're talking about in class that day, or at least those you've been discussing recently. Been talking about graphing lines in the coordinate plate? The SAT *loves* graphing lines in the coordinate plane! Here, find this slope.

Of course, it's not always that easy. Sometimes it's tough to find a problem that exactly matches your current topic. Maybe you can find one, but not a whole week's worth. Again, there aren't that many problems out there. So if you want to commit to doing this on a regular basis, at some point you'll likely have to go off-topic. That's okay. Just keep a few things in mind.

- As we've mentioned, don't do problems that are wildly above their skill level. If they're just starting algebra 1, don't give them a problem where you have to divide polynomials and find the remainder.
- That said, it's sometimes nice to do problems involving new concepts if you can use that problem to introduce the concept. For example, say there's a problem dealing with an exponential function, which you haven't talked about yet. But the problem doesn't require complex computation or graphing, it just requires you to understand the meaning of the components of the equation. You can take this opportunity to show them the equation and explain what the parts mean. They likely won't get it right today, but many of them will the next time they see an exponential function, now that it's familiar to them.
- You may also decide to target your work based on data from test results. Whether that's from a practice test you administer yourself or from the real PSAT, if you give a test, you're likely to have an abundance of information about your students' performance. Look for areas where they need the most help and come up with some targeted drills on those subjects.
- Be aware of the difficulty level of the problem. If it's a hard question, you may want to give students more time than usual (and it may take you longer to go over it). It's okay to give a hard question to a remedial class; we firmly believe that anyone is capable of getting any question if given the right training and time constraints. But if it's a very hard question *and* it deals with a concept you haven't covered recently, you may want to save it until it's more directly relevant to your classwork. Note that while questions on practice tests are numbered by difficulty, questions acquired from other sources may not have difficulty information, so you'll have to use your judgment.

Where Can I Get Problems?

The best place to get problems is straight from the College Board. As of the first administration of the new test in March 2016, their website has four full practice SATs available for free download, one full PSAT, and two sets of math practice problems. They will also subsequently release one of the real spring 2016 tests for free download by June.

But, of course, you may not want to use the practice tests for daily classwork if you're saving them to be administered as proctored tests. Here are a few places where you can find additional practice material:

- The College Board offers an SAT Question of the Day service on their website. These are good questions, delivered daily. But it's potluck what the topic is going to be, and it's not always a math question.
- As mentioned earlier, the College Board has partnered with Khan Academy to provide additional practice material free of charge at khanacademy.org. There's a *lot* of material here, thousands of questions, organized by topic and difficulty. It is meant to be an online service, so it is difficult to print the problems and can be tough to navigate if you're looking for a particular problem. It's easy enough to project the problems if you have a Smart Board or similar device, but the real test is still given on paper. Writing directly on a passage or figure is an important way to eliminate careless mistakes, and that's not possible when doing questions online.
- A-List's SAT book, *The Book of Knowledge*, is a thorough examination of the content of the new test and contains drills separated by content area along with lectures and exercises to help do the problems effectively. A-List also has an online platform that offers a suite of supplemental material like quizzes and worksheets.
- If you're really in a crunch, you could even use some old-format SATs. The format and timing have changed dramatically, but most of the math problems from the old test are good math problems that are still relevant to the new test. New content was added to the redesigned test, but for the most part very few concepts were removed. So just about any question from the old test will be very familiar. (The other major difference is that the old test had five choices for the multiple-choice questions, not four like the redesigned test has.)

Translation Exercises

What if you want some middle ground? You're not giving a prep course and don't have time for full tests and sections, but you want to do more serious work in class than just a question a day. That's when you're going to want to have content-specific drills. You can find such drills in *The Book of Knowledge* or on the Khan Academy website. But you can also make your own drills using existing test material.

This is a process we call *translation*. Take an existing problem as a framework and rewrite it to make a new problem. This can be as simple as just changing the numbers or changing the setting. But you can also adjust it as you see fit. Add more concepts, trim it to fewer concepts, make it harder, make it easier, do what you want with it.

This is a valuable exercise not only because the output is a new problem set you can use, but also because it forces you to analyze the problem deeply to understand the root concepts. It forces you to think about not just the correct answer but also the incorrect answers. How will your students see this problem? Think about not just your solution but also your students' solutions.

There are several different ways you can go about doing this.

Direct Translation

You can translate questions to greater or lesser degrees, depending on your goal. Let's start with a simple problem:

(Note: The following problems are numbered sequentially for convenience. The numbers do *not* indicate difficulty level.)

1. **Bob has 4 dollars more than Lisa. If Lisa has x dollars, how much would Bob have if he doubled his money?**
   A) $x + 4$
   B) $x + 8$
   C) $2x + 4$
   D) $2x + 8$

Let's ignore the choices for now and focus on the question. The easiest way to change the question is just to change the names:

2. **Sandeep has 4 dollars more than Carrie. If Carrie has d dollars, how much would Sandeep have if he doubled his money?**

This is hardly any change at all. It's basically identical to the original problem. (Note: the College Board itself is no stranger to translation. The last version of *The Official SAT Study Guide* for the old test contained a few questions that were cursory translations of each other, like this.) Let's not make it about money. Let's make it about something else.

3. **Alan has 4 more cats than Lane has. If Lane has *c* cats, how many would Alan have if he doubled his cats?**

Getting a little crazy, but okay. We're still talking about people owning objects, though. Let's try a different direction:

4. **The Rockets won 3 more games than the Spurs. If the Spurs won *z* games, how many games would the Rockets win if they won twice as many games?**

This starts out good, but ends up nonsensical. Moving into the realm of "wins" changes the problem conceptually to intangible objects. But, as we can see, these intangible objects don't behave in quite the same way. Look at that: how many games would they win if they won twice as many games? What does that mean? We kept the underlying math the same, but we need to change the wording. We've got to be more explicit about the passage of time. Let's try again:

5. **In 2014, the Spurs won z games, and the Rockets won 4 more games than the Spurs. In 2015, the Rockets won twice as many games as they did in 2014. In terms of z, how many games did the Rockets win in 2015?**

Great. Note that mathematically, question 5 is *exactly* the same as question 1. The only difference is the letter used for the variables and the concepts they represent. But do you think the question will be perceived the same way by your students? Which version do you think is easier or harder? Or are they both the same? I would argue that question 5 is probably harder than question 1 because of the increased abstraction of the concepts.

As math teachers, it can sometimes be tempting to treat questions with identical underlying math as identical problems, but students don't always see it that way. We've all seen students who can perform well on pure equations (which can often be done through rote memorization of processes and muscle memory) but perform poorly when the same concepts appear in a context.

This is a particularly important issue if you're teaching nonnative English speakers. English language learners (ELL) may be excellent math students but struggle to decipher long English sentences. It can even be an issue for students who are native speakers but are just poor readers. For them, the SAT is twice as much work. And twice as much work for you, too. First, you have to teach them the underlying math concepts they need to get the question. Then, you have to teach them how to identify those concepts based on the question as it's written.

Analysis in Science, Analysis in History/Social Studies

This issue of the topic of the problems is relevant because it's reported on students' score reports. As we mentioned earlier, students will get two "cross-test scores" in *Analysis in Science* and *Analysis in History/ Social Studies*. These categories take questions from all sections of the test: Reading, Writing, and Math.

Reading and Writing are both passage-based sections, so it's clear which questions will be used for these cross-test scores. There will be passages about science and there will be passages about history. Math questions work the same way. You'll see some questions in science-related settings (like a function that describes a scientific relationship or data collected from the field), and others in history/social science–related settings (most of these relate to money and economic issues, but some also deal with demographic data).

In practice, however, the SAT is inconsistent in the way it applies these tags. All questions assigned to these categories seem to be appropriately assigned—we haven't found any instances of a question tagged as science that didn't seem to be legitimately about science. But there are a large number of questions that didn't get assigned to either category but easily could have. For example:

- On test 1, section 4, questions 9 and 10 (1.4.9 and 1.4.10 in our code system) both refer to the same equation describing the relationship between air temperature and the speed of sound, but question 9 was tagged as science and question 10 was not.
- Question 4.3.12, about writing algebraic expressions to calculate a tip at a certain percentage, was tagged as history/social studies. Question 1.4.20, about writing algebraic expressions to calculate the price of a laptop after certain percentages of discount and tax are applied, did not receive a tag.

- Question 3.4.25, about calculating the volume of a cylindrical silo, was tagged as science, but question 1.4.35, about calculating the volume of a cylindrical silo, did not receive a tag.

There are many more examples of questions that appear in a history or science context but are not included in the cross-test scores (such as 3.4.24 or 3.3.4). For Reading and Writing, the College Board specified which questions would contribute to the cross-test scores: for Reading, all questions in passages of the relevant subject type; for Writing, only the Expression of Ideas question in the relevant passages. But they have not stated specifically how they determine which math questions go into the cross-test scores, only that eight math questions will contribute to each score (although even that is wrong: practice test 2 had only seven questions in the Science category).

It seems obvious that these cross-test scores only use a sample of the math questions that use the specified contexts. Why they would do that is unclear. There does not seem to be any strong pattern for which questions get included and which do not. Perhaps the College Board wanted to have a large number of questions in real-world settings (which they have said explicitly in their discussion of the redesign), but wanted the cross-test score to be more heavily weighted to the Reading questions.

The value of these cross-test scores is already suspect, because the College Board has provided no evidence that the ability to comprehend a science passage is at all related to the ability to do a math problem. As we've seen in our translations above, problems can have the exact same underlying math but be put in different contexts. To wit, compare this problem:

6. **Sherwyn deposits $10,000 in a bank account that gets 5% annual compound interest. Which of the following equations describes the amount of money, *C*, Sherwyn will have in his account after *t* years?**

. . . with this problem:

7. **Sherwyn determines that a certain region contains 10,000 frogs. If the frog population grows by 5% each year, which of the following equations describes the number of frogs, *F*, in the region after *t* years?**

These problems contain exactly the same mathematical principles, but one is in an economic context and the other in a scientific context. We could even remove the context altogether:

8. **Which of the following equations shows an exponential function with 5% growth and an initial value of 10,000?**

Is it possible that some students are better at one context than another? Sure, absolutely. But does that tell us anything meaningful about those students' abilities?

All these issues teach us a few important lessons:

- Context is irrelevant to mathematical content. The same content can be provided in different contexts.
- Context can affect the difficulty of a question, in particular by making things seem more abstract or concrete.
- Some students will struggle with certain contexts due to individual preferences or skills. Some will struggle with all contexts due to poor reading comprehension.
- Students' performance in different contexts may be significant but tells us little about their understanding of the underlying mathematical principles.
- We should strive to get students to see the irrelevance of context, to ignore context and focus on the math.
- We can demonstrate the malleability of context by taking any problem and putting it in a new context or stripping the context away altogether.
- Don't pay attention to the cross-test scores.

Distractor Choices

Let's go back to our first sample question:

1. **Bob has 4 dollars more than Lisa. If Lisa has x dollars, how much would Bob have if he doubled his money?**
   A) $x + 4$
   B) $x + 8$
   C) $2x + 4$
   D) $2x + 8$

How do we do this question? We want to take these sentences in English and rewrite them as algebraic expressions.

- Lisa has x dollars.
- Bob has 4 more, so Bob has $x + 4$ dollars.
- If Bob doubled his money he'd have $2(x + 4)$.
- Distribute across the parentheses to get $2x + 8$. That's choice D).

Let's look at the wrong answers here. What do you think the most popular wrong answer would be? It's probably choice C), $2x + 4$. You get that if you forget to distribute the 2 inside the parentheses. This is particularly common if you don't write down your work and only think about it in your head, since we don't tend to pronounce parentheses:

> *Bob has x plus four, so if he doubled his money he'd have two times x plus four. Ah, choice C) says two times x plus four. I am finished.*

What about the other choices? Choice A), $x + 4$, is the amount Bob has now, before he doubles his money. Choice B), $x + 8$, has the opposite problem as choice C): you doubled the 4 but not the x.

We call wrong answers *distractor choices*. Strictly speaking, the term "distractor" refers to any wrong answer choice in a problem, since they distract you from the right answer. But some wrong choices are more distracting than others. The SAT loves to plant incorrect choices that aren't random but are answers students would get by making common mistakes. This seems evil to students, and, well, it's not not-evil, but it is important information for you because it makes it easier for you to diagnose their mistakes. If you know a student picked C) on this question, you know exactly what they did wrong before they even say anything. A-List offers an online remote test-grading service that provides reports with answer-choice-level information for your class. These reports show the percentage of students who chose each choice for each question. This way, you can immediately see not only which questions most students got wrong, but also which wrong answers were most popular.

Not every choice will be equally tempting, of course. Some wrong answers really are just random numbers that only students who pick randomly will choose. Others are so tempting that their mere presence is the only thing that makes a hard question hard: the underlying math is easy, but *everybody* looks at C) and picks it right away, without thinking.

You should explicitly talk about distractor choices when reviewing real tests, especially if you see one that's particularly tempting. Don't just give the right answer and stop. Look at the wrongs and ask: why might someone think it's A)? Why is A) not the answer? When these ideas are out in the open, students will be prepared to face them.

When you're translating a question, you'll want to retain the distractors that are already there, but you may have to adapt them for any new numbers you picked. For example, let's look at the new version of question 1 that we came up with, now with answer choices.

5. **In 2014, the Spurs won z games, and the Rockets won 4 more games than the Spurs. In 2015, the Rockets won twice as many games as they did in 2014. In terms of z, how many games did the Rockets win in 2015?**
   A) $z + 4$
   B) $z + 8$
   C) $2z + 4$
   D) $2z + 8$

Here we have all the same numbers as the original, so our choices should be exactly the same, just using z instead of x. But let's say we changed the 4 to a 10. We'd change the correct answer to $2x + 20$, of course, but we'd also want to change the wrong answers analogously:

9. **In 2014, the Spurs won z games, and the Rockets won 10 more games than the Spurs. In 2015, the Rockets won twice as many games as they did in 2014. In terms of z, how many games did the Rockets win in 2015?**
   A) $z + 10$
   B) $z + 20$
   C) $2z + 10$
   D) $2z + 20$

Now let's say we want to stray further from the original. Let's make the Rockets worse in 2015:

10. **In 2014, the Spurs won z games, and the Rockets won 10 more games than the Spurs. In 2015, the Rockets won <u>half</u>**

as many games as they did in 2014. In terms of z, how many games did the Rockets win in 2015?

Now instead of $2(z + 10)$, the correct answer should be $\frac{z+10}{2}$. Do we want the answer to appear like that? Or $\frac{z}{2} + 5$? Or $\frac{1}{2}z + 5$? Let's go with the last option to make it most analogous to the original. We can fill out the rest of the choices the same way:

10. **In 2014, the Spurs won z games, and the Rockets won 10 more games than the Spurs. In 2015, the Rockets won <u>half</u> as many games as they did in 2014. In terms of z, how many games did the Rockets win in 2015?**
    A) $z + 5$
    B) $z + 10$
    C) $\frac{1}{2}z + 5$
    D) $\frac{1}{2}z + 10$

First, we should note that the answer is now C) instead of D). Choices are generally arranged in order from smallest to largest when they give values. When the choices are expressions like this, there's a bit more leeway about the order, but they're still grouped in some logical way.

But now we should think: maybe in this new version students won't make exactly the same mistakes. In the original, kids might mess up how to apply the 2, but they'll all correctly *multiply* the 2. Fractions confuse students, and it's certainly possible students will get confused about which direction to go after they get to $z + 10$. So let's add a distractor where they multiply by 2 instead of dividing by 2:

11. **In 2014, the Spurs won z games, and the Rockets won 10 more games than the Spurs. In 2015, the Rockets won <u>half</u> as many games as they did in 2014. In terms of z, how many games did the Rockets win in 2015?**
    A) $\frac{1}{2}z + 5$
    B) $\frac{1}{2}z + 10$
    C) $z + 5$
    D) $2x + 20$

Since we only get four choices, we had to drop an old one in order to add a new one. Sad, isn't it? Sometimes we come up with more good distractor choices than we have room for.

Grid-Ins

Of course, another alternative is to drop answer choices altogether. Remember that the test has 13 grid-in questions where students must provide their own answers. These problems are generally no different than regular multiple-choice problems, so they don't require special skills, but they can be tough for students who like to eliminate.

Here are some general points about grid-in answers.

- There are only four spaces in the grid where you put your answer. You don't have to use all the spaces, but our answer can't be a five-digit number.
- Fractions and decimals are permitted as answers. If an answer contains a decimal point or a fraction bar, that symbol counts as one of the four spaces. So the answer cannot be 11/13, because that would use five spaces.
- There is no negative symbol on the grid. Negative answers are never permitted on grid-ins.
- Repeating decimals are permitted as long as you grid in as many spaces as will fit. If the answer is 1/3, you will get credit for .333 but not 0.33. If the answer is 4/3, you will get credit for 1.33 but not 1.3. If the answer is 2/3 you can grid in .666 or .667 but not 0.66. (Of course, you can always grid in the fractions, too.)
- Mixed numbers must be converted to improper fractions. If you try to grid in $1\frac{1}{3}$, it will be read as "11/3". Put down 4/3 instead.
- You do not have to reduce fractions to the lowest term. If the answer is 1/2, you will get credit for 2/4. But not for 10/20, since that won't fit in the grid.

Of course, this isn't the real SAT, this is your drill, so it's up to you if you want to follow these rules. If you want your answer to be −15,763, be my guest. Just realize that your question will be noncanonical.

Obviously, we can't just use our newly translated question as a grid-in because our answer is an expression; the answer to a grid-in must be a number. So if we want a grid-in, we'll have to adapt it.

12. In 2014, the Rockets won 10 more games than the Spurs. In 2015, the Rockets won half as many games as they did in 2014. If the Spurs won 40 games in 2014, how many games did the Rockets win in 2015?

The Rockets won 10 + 40 in 2014—that's 50 games. They won half as many in 2015—that's 25 games.

Do you think question 12 is now easier or harder than question 11 was? It's probably easier because we're dealing with concrete numbers. There's no danger of misapplying the parentheses because there are no parentheses.

This, by the way, is why we advocate a technique called **Plug In** for questions like number 11. Pick a number for z to get an answer for the number of games. Then put your number in the choices for z and see which gives you the same answer. You can do this for just about any question that has variables in the answer choices. If the answer is $x + 4$, it's *always* $x + 4$, regardless of what the value of x is. So pick one and try it.

If the problem is too easy for you now, we can make it harder by using harder numbers:

13. In 2014, the Rockets won 5 fewer games than the Spurs. In 2015, the Rockets won 7/8 as many games as they did in 2014. If the Spurs won 61 games in 2014, how many games did the Rockets win in 2015?

Ugh. Okay, fine. 61 – 5 = 56 wins in 2014. (7/8) × 56 = 49 wins in 2015.

That one was still too easy because we had to stick to integers since we're talking about games. Let's try going back to the original version:

14. Bob has $3.80 more than Lisa. If Lisa has $2.18, how much would Bob have if he doubled his money?

Now it's harder, sure. But the difficulty here is purely computational. If students have calculators, it's a snap. Question 13 is probably a bit harder because students have to deal with the fraction, 7/8.

Adapted Translation

We've seen ways for you to alter existing problems while retaining the original concepts in the problem. This is useful for several reasons:

- To demonstrate the irrelevance of context to the underlying math.

- To create more problems in particular contexts that are difficult or troublesome for your students.
- To create more problems dealing with a specific concept.

But who says you have to stick to the original concepts? We've already seen how we can adapt problems beyond their original scope. While we're at it, let's keep going and add more concepts.

Let's start with our original problem:

1. **Bob has 4 dollars more than Lisa. If Lisa has x dollars, how much would Bob have if he doubled his money?**
   A) $x + 4$
   B) $x + 8$
   C) $2x + 4$
   D) $2x + 8$

What's something that students struggle with? How about percent change?

15. **Bob's salary is 30% greater than Lisa's.**

Wait, stop, no. Let's not make the man's salary higher than the woman's. Start again:

15. **Molly's salary is 30% greater than Noah's. If Noah's salary is x dollars, which of the following gives Molly's salary after she gets a 20% raise?**

Our previous translations were variations on the same problem: the numbers were different but the concepts were the same. This problem is now significantly different from the original because it introduced a new concept of percent change.

Let's do the problem first:

- Noah gets x dollars.
- Molly gets 30% more, so she gets $1.3x$
- She gets a 15% raise over what she had, so that's $1.2(1.3x)$
- Distribute across the parentheses to get $\mathbf{1.56x}$

Let's call that choice A). We could leave it as a decimal like that. Or use $\left(\frac{156}{100}\right)x$. Sometimes SAT problems will be written to make the math steps apparent, so you could make the choice $(1.2)(1.3)x$, or even $x + 0.56x$.

Here are some possible distractor choices:

B) 0.06x That's just 30% times 20%.
C) 0.26x That's 20% of 130%.
D) 0.36x That's 30% of 120%.
E) 1.3x That's her starting salary.
F) 1.5x That's adding 30% and 20%.

All of these distractors assume students know that students know to multiply percentages. But maybe they don't! We could have distractors like these:

G) 30x + 20

H) $\frac{x}{1.56}$

I) $\frac{1.3x}{1.2}$

Now we're just shuffling these numbers around randomly, but such things can still be tempting for students who have no idea how to proceed. In general, the more tempting the distractor choices, the harder the question is. Some students may get choice C) above just because they're careless. If a student gets that as an answer and sees that it's a choice, they'll pick it and move on without thinking. But if it's *not* a choice, they'll know they made a mistake and are more likely to go back and check their work. So having *random* distractors doesn't guarantee that more students will *understand* a question, but it makes it less likely they'll make *careless* mistakes.

Let's see a variation on this question in order to demonstrate that.

16. Molly's salary is 30% greater than Noah's. If Molly gets a 20% raise, her salary is now what percent greater than Noah's?
A) 6%
B) 10%
C) 50%
D) 56%

Every single student is going to pick choice C) here. Every single one. It is so tempting to just add 30 and 20 to get 50. So when they see that 50 there, their instincts are confirmed and they move on to the next problem. But if 50 wasn't there, if instead our choices were

A) 26%
B) 36%
C) 46%
D) 56%

. . . then they'd know something was up. That doesn't mean they'd necessarily know the *right* way to do the problem, but they'd know their instinct was wrong. They'd spend more time thinking about it and on average more students would get it right.

Let's keep adding concepts. The redesigned SAT loves systems of equations. Let's add another equation here.

16. If Molly's salary is 30% greater than Noah's and their combined salary is $92,000, what is the value of Molly's salary?

Now we have two equations

$m = 1.3n$
$m + n = 92{,}000$

We can substitute to get

$1.3n + n = 92{,}000$
$2.3n = 92{,}000$
$n = 40{,}000$ That's Noah's salary. Subtract from 92,000 to get
$m = 52{,}000$

Let's look at those numbers we used to get our answer. Do any of those look like good numbers to use as a distractor choices? YES. **40,000.** *You should absolutely include 40,000 as a distractor choice.* If you do, half your students will pick it. They will go through all the correct steps, get a value for one of the variables, then assume that must be the answer without actually checking if that's the variable they want.

This is a great method for coming up with distractor choices. Go through all the correct math you've done and look for numbers you used that weren't the correct answer. Use those as choices. Somebody will get that number and stop.

Another great distractor idea: just take 30% of $92,000. That's $27,600. Great, now subtract that from $92,000. That's $64,400. Perfect, we've got four answer choices now:

16. If Molly's salary is 30% greater than Noah's and their combined salary is $92,000, what is the value of Molly's salary?

A) $27,600
B) $40,000
C) $52,000
D) $64,400

Now that we have answer choices, this question is a great demonstration of our other math technique, **Backsolve**. Pick an answer choice and put it into the problem to see if the numbers work out. Try choice B), $40,000. If that's the answer, then that's Molly's salary. So $40,000 is 30% higher than Noah's. Divide by 1.3 to get Noah's salary, $30,769.23.

If this is right, the two salaries will add up to $92,000. They obviously don't; they only add up to around $70,000. So B) was too small. Try the same with choice C) and it will work out.

The nice thing about Backsolve is you start directly from the question, so you're much less likely to choose the wrong variable. The whole first step is: let's make the choice the answer to the question. You still need to know how to manipulate percentages, but you're less likely to make a careless error.

We can see now that adapted translation is a great way to make your own questions when you want drills on specific concepts. Want to work on systems of equations? Find questions with one equation and add another parameter. Want to work on graphing lines? Find questions with ungraphed linear functions and ask students to graph them. Do both of these things together: add an equation and graph them both. Look: we started with one fairly easy problem and made 16 versions of it. And we didn't even add that many concepts to it. The only limits are your imagination.

Student Work

That seems great and all, but that seems like a lot of work. How can we get drills to the students faster?

Why not have the students themselves do the work?

This may seem like a cheap way for you to get out of work, but it isn't (entirely). The discussion above shows that even simple problems can spur long discussions of what is being tested and how to go about solving problems. These are conversations that we can be having with our students explicitly.

The issue of distractor choices, for example, shouldn't be a secret. It's important that they know about them, that the people who write the questions *intentionally* make wrong answers available that are the result of predictable errors. Give your students these instructions:

1. **Pick a problem.** Or, assign problems to the students.
2. **Translate the problem** into a new problem. You can give specific parameters for what the problem looks like. For example:
   a. Be sure the translated problem contains none of the same numbers as the original. (Or set limits on how many numbers they're allowed to keep.)
   b. Be sure the translated problem is in a different context. Don't just change basketball teams to baseball teams; make it significantly different. You can specify the context if you like (e.g., make it a science question).
3. **Write distractor choices.** Be sure at least two incorrect choices have specific incorrect paths that lead towards them.
4. **Write answer explanations for the question.** This is a great tool. Nothing gets students to dive into a problem like being forced to explain it to another student. Be sure to:
   a. Include two ways of solving the problem. They don't have to be equally good ways—one might be clearly faster or easier than the other—but they both have to be mathematically valid ways.
   b. Explain why students might pick the distractor choices. What mistakes would lead to those choices?
5. **Analyze how the problem has changed.** Has the translation made the problem easier or harder? Why?

In reviewing the students' work, there will likely be some translations that don't work. The original problem was fine, but the translation is problematic or outright nonsensical. See, for example, our awkward wording for question 4 above. If it doesn't work, talk about it. *Why* didn't it work? What did the student do to the problem that made it problematic?

Be sure also to check that the students' wrong answers are all actually wrong! Sometimes students will accidentally make a wrong answer right in their efforts to be sneaky. For example, they might give two choices with algebraic expressions that seem distinct but are actually equivalent, like $2x + 8$ and $2(x + 4)$.

This is also a great exercise to do in groups. Have every student write a translation, then pass it to the left so another writes the distractor choices. Then pass it down once more to have a third write the answer explanations. When they're done, they'll have their own drill. Have groups swap drills and do each other's questions. See how they do!

Translation Summary

- Pick an SAT problem. Any problem.
  - Analyze the concepts involved in the problem. What rules and skills do you need here?
  - Analyze the answer choices. Which ones are tempting? Why might a student pick one?
- Write a new problem using the *same* skills in a new context.
  - Try writing the same problem in other settings. Try science or history/social studies. Try writing it with no setting at all.
  - Did your problem come out easier, harder, or the same? Why?
  - Use new numbers. Try coming up with new distractor choices.
  - Could it be a grid-in? How would that change the difficulty?
- Keep the same framework as the original, but *add concepts* to the problem. E.g., if it was one equation translation, make it a system. Add percentages. Add the coordinate plane.
  - How would that alter the difficulty?
  - Could you use a math technique on the original problem? Can you use one on your new problem? Try to adapt the original into a Plug In or a Backsolve problem.
  - Add distractor choices to fit the new content. Find distractors by looking at the math work you did in order to get the right answer. Use some of your intermediary steps as wrong choices.

Appendix
All Math Alignment Tables

How to Read the Tables

Each table refers to a single conceptual category. Within each table, standards are grouped by domain and cluster. Remember that standards are numbered continuously within each domain, irrespective of clusters. Each table contains columns showing:

- The code for the standard, as defined by the CCSSI
- The standard itself
- Its alignment with the SAT

Each standard has a code containing three parts: a letter denoting the conceptual category, a series of letters denoting the domain, and a number denoting the standard. If the standard has any subskills, the skills will be denoted by a letter following the standard number. **Code: [category]-[domain].[standard][subskill]**

For example, "F-IF.7c" denotes: the Functions category [F], the "Interpreting Functions" domain [IF], the seventh standard within that domain [7], and the third subskill associated with that standard [c].

The two alignment columns will each display one of the following symbols:

- **Y** = The standard is aligned with the test in question.
- **N** = The standard is not aligned with the test in question.
- **P** = The standard is partially aligned with the test in question. This means a qualifying comment was listed for the standard in the original alignment document.

Standards marked with * are modeling standards. Standards marked with (+) are standards beyond college and career readiness.

Code	Standards for Mathematical Practice [MP]	SAT
MP.1	1. Make sense of problems and persevere in solving them.	**Y**
MP.2	2. Reason abstractly and quantitatively.	**Y**
MP.3	3. Construct viable arguments and critique the reasoning of others.	**Y**
MP.4	4. Model with mathematics.	**Y**
MP.5	5. Use appropriate tools strategically.	**Y**
MP.6	6. Attend to precision.	**Y**
MP.7	7. Look for and make use of structure.	**Y**
MP.8	8. Look for and express regularity in repeated reasoning.	**Y**

Code	Standard	SAT
Numbers and Quantity		
The Real Number System [N-RN]		
Extend the properties of exponents to rational exponents.		
N-RN.1	1. Explain how the definition of the meaning of rational exponents follows from extending the properties of integer exponents to those values, allowing for a notation for radicals in terms of rational exponents. *For example, we define* $5^{1/3}$ *to be the cube root of 5 because we want* $(5^{1/3})^3 = 5^{(1/3)3}$ *to hold, so* $(5^{1/3})^3$ *must equal 5.*	**Y**
N-RN.2	2. Rewrite expressions involving radicals and rational exponents using the properties of exponents.	**Y**
Use properties of rational and irrational numbers.		
N-RN.3	3. Explain why the sum or product of two rational numbers is rational; that the sum of a rational number and an irrational number is irrational; and that the product of a nonzero rational number and an irrational number is irrational.	**Y**
Quantities* [N-Q]		
Reason quantitatively and use units to solve problems.		
N-Q.1	1. Use units as a way to understand problems and to guide the solution of multistep problems; choose and interpret units consistently in formulas; choose and interpret the scale and the origin in graphs and data displays.	**Y**
N-Q.2	2. Define appropriate quantities for the purpose of descriptive modeling.	**Y**
N-Q.3	3. Choose a level of accuracy appropriate to limitations on measurement when reporting quantities.	**Y**
The Complex Number System [N-CN]		
Perform arithmetic operations with complex numbers.		
N-CN.1	1. Know there is a complex number i such that $i^2 = -1$, and every complex number has the form $a + bi$ with a and b real.	**Y**
N-CN.2	2. Use the relation $i^2 = -1$ and the commutative, associative, and distributive properties to add, subtract, and multiply complex numbers.	**Y**
N-CN.3	3. (+) Find the conjugate of a complex number; use conjugates to find moduli and quotients of complex numbers.	**N**

(Continued)

Code	Standard	SAT
Represent complex numbers and their operations on the complex plane.		
N-CN.4	4. (+) Represent complex numbers on the complex plane in rectangular and polar form (including real and imaginary numbers), and explain why the rectangular and polar forms of a given complex number represent the same number.	**N**
N-CN.5	5. (+) Represent addition, subtraction, multiplication, and conjugation of complex numbers geometrically on the complex plane; use properties of this representation for computation. *For example,* $(-1 + \sqrt{3}i)^3 = 8$ *because* $(-1 + \sqrt{3}i)$ *has modulus 2 and argument 120°.*	**N**
N-CN.6	6. (+) Calculate the distance between numbers in the complex plane as the modulus of the difference, and the midpoint of a segment as the average of the numbers at its endpoints.	**N**
Use complex numbers in polynomial identities and equations.		
N-CN.7	7. Solve quadratic equations with real coefficients that have complex solutions.	**Y**
N-CN.8	8. (+) Extend polynomial identities to the complex numbers. *For example, rewrite* $x^2 + 4$ *as* $(x + 2i)(x - 2i)$.	**Y**
N-CN.9	9. (+) Know the Fundamental Theorem of Algebra; show that it is true for quadratic polynomials.	**Y**
Vector and Matrix Quantities [N-VM]		
Represent and model with vector quantities.		
N-VM.1	1. (+) Recognize vector quantities as having both magnitude and direction. Represent vector quantities by directed line segments, and use appropriate symbols for vectors and their magnitudes (e.g., $\boldsymbol{v}$, $\|\boldsymbol{v}\|$, $\|\|\boldsymbol{v}\|\|$, v).	**N**
N-VM.2	2. (+) Find the components of a vector by subtracting the coordinates of an initial point from the coordinates of a terminal point.	**N**
N-VM.3	3. (+) Solve problems involving velocity and other quantities that can be represented by vectors.	**Y**

Code	Standard	SAT
Perform operations on vectors.		
N-VM.4	4. (+) Add and subtract vectors.	—
N-VM.4a	a. Add vectors end-to-end, component-wise, and by the parallelogram rule. Understand that the magnitude of a sum of two vectors is typically not the sum of the magnitudes.	**N**
N-VM.4b	b. Given two vectors in magnitude and direction form, determine the magnitude and direction of their sum.	**N**
N-VM.4c	c. Understand vector subtraction $\boldsymbol{v} - \boldsymbol{w}$ as $\boldsymbol{v} + (-\boldsymbol{w})$, where $-\boldsymbol{w}$ is the additive inverse of $\boldsymbol{w}$, with the same magnitude as $\boldsymbol{w}$ and pointing in the opposite direction. Represent vector subtraction graphically by connecting the tips in the appropriate order, and perform vector subtraction component-wise.	**N**
N-VM.5	5. (+) Multiply a vector by a scalar.	
N-VM.5a	a. Represent scalar multiplication graphically by scaling vectors and possibly reversing their direction; perform scalar multiplication component-wise, e.g., as $c(vx, vy) = (cvx, cvy)$.	**N**
N-VM.5b	b. Compute the magnitude of a scalar multiple $c\boldsymbol{v}$ using $\|\|c\boldsymbol{v}\|\| = \|c\|v$. Compute the direction of $c\boldsymbol{v}$ knowing that when $\|c\|v \neq 0$, the direction of $c\boldsymbol{v}$ is either along $\boldsymbol{v}$ (for $c > 0$) or against $\boldsymbol{v}$ (for $c < 0$).	**N**
Perform operations on matrices and use matrices in applications.		
N-VM.6	6. (+) Use matrices to represent and manipulate data, e.g., to represent payoffs or incidence relationships in a network.	**N**
N-VM.7	7. (+) Multiply matrices by scalars to produce new matrices, e.g., as when all of the payoffs in a game are doubled.	**N**
N-VM.8	8. (+) Add, subtract, and multiply matrices of appropriate dimensions.	**N**
N-VM.9	9. (+) Understand that, unlike multiplication of numbers, matrix multiplication for square matrices is not a commutative operation, but still satisfies the associative and distributive properties.	**N**

(Continued)

Code	Standard	SAT
N-VM.10	10. (+) Understand that the zero and identity matrices play a role in matrix addition and multiplication similar to the role of 0 and 1 in the real numbers. The determinant of a square matrix is nonzero if and only if the matrix has a multiplicative inverse.	**N**
N-VM.11	11. (+) Multiply a vector (regarded as a matrix with one column) by a matrix of suitable dimensions to produce another vector. Work with matrices as transformations of vectors.	**N**
N-VM.12	12. (+) Work with 2×2 matrices as transformations of the plane, and interpret the absolute value of the determinant in terms of area.	**N**

* indicates modeling standard (+) indicates standard beyond college and career readiness

Code	Standard	SAT
Seeing Structure in Expressions [A-SSE]		
Interpret the structure of expressions.		
A-SSE.1	1. Interpret expressions that represent a quantity in terms of its context.*	—
A-SSE.1a	a. Interpret parts of an expression, such as terms, factors, and coefficients.	**Y**
A-SSE.1b	b. Interpret complicated expressions by viewing one or more of their parts as a single entity. *For example, interpret $P(1 + r)^n$ as the product of P and a factor not depending on P.*	**Y**
A-SSE.2	2. Use the structure of an expression to identify ways to rewrite it. *For example, see $x^4 - y^4$ as $(x^2)^2 - (y^2)^2$, thus recognizing it as a difference of squares that can be factored as $(x^2 - y^2)(x^2 + y^2)$.*	**Y**
Write expressions in equivalent forms to solve problems.		
A-SSE.3	3. Choose and produce an equivalent form of an expression to reveal and explain properties of the quantity represented by the expression.*	—
A-SSE.3a	a. Factor a quadratic expression to reveal the zeros of the function it defines.	**Y**
A-SSE.3b	b. Complete the square in a quadratic expression to reveal the maximum or minimum value of the function it defines.	**Y**
A-SSE.3c	c. Use the properties of exponents to transform expressions for exponential functions. *For example, the expression 1.15^t can be rewritten as $(1.15^{1/12})^{12t} \approx 1.012^{12t}$ to reveal the approximate equivalent monthly interest rate if the annual rate is 15%.*	**Y**
A-SSE.4	4. Derive the formula for the sum of a finite geometric series (when the common ratio is not 1), and use the formula to solve problems. *For example, calculate mortgage payments.**	**P**
Arithmetic with Polynomials and Rational Expressions [A-APR]		
Perform arithmetic operations on polynomials.		
A-APR.1	1. Understand that polynomials form a system analogous to the integers, namely, they are closed under the operations of addition, subtraction, and multiplication; add, subtract, and multiply polynomials.	**Y**
Understand the relationship between zeros and factors of polynomials.		
A-APR.2	2. Know and apply the Remainder Theorem: For a polynomial $p(x)$ and a number a, the remainder on division by $x - a$ is $p(a)$, so $p(a) = 0$ if and only if $(x - a)$ is a factor of $p(x)$.	**Y**
A-APR.3	3. Identify zeros of polynomials when suitable factorizations are available, and use the zeros to construct a rough graph of the function defined by the polynomial.	**Y**

(Continued)

Code	Standard	SAT
Use polynomial identities to solve problems.		
A-APR.4	4. Prove polynomial identities and use them to describe numerical relationships. *For example, the polynomial identity* $(x^2 + y^2)^2 = (x^2 - y^2)^2 + (2xy)^2$ *can be used to generate Pythagorean triples.*	**Y**
A-APR.5	5. (+) Know and apply the Binomial Theorem for the expansion of $(x + y)^n$ in powers of x and y for a positive integer n, where x and y are any numbers, with coefficients determined for example by Pascal's Triangle.	**Y**
Rewrite rational expressions.		
A-APR.6	6. Rewrite simple rational expressions in different forms; write $a(x)/b(x)$ in the form $q(x) + r(x)/b(x)$, where $a(x)$, $b(x)$, $q(x)$, and $r(x)$ are polynomials with the degree of $r(x)$ less than the degree of $b(x)$, using inspection, long division, or, for the more complicated examples, a computer algebra system.	**Y**
A-APR.7	7. (+) Understand that rational expressions form a system analogous to the rational numbers, closed under addition, subtraction, multiplication, and division by a nonzero rational expression; add, subtract, multiply, and divide rational expressions.	**Y**
Creating Equations [A-CED]		
Create equations that describe numbers or relationships.		
A-CED.1	1. Create equations and inequalities in one variable and use them to solve problems. *Include equations arising from linear and quadratic functions, and simple rational and exponential functions.*	**Y**
A-CED.2	2. Create equations in two or more variables to represent relationships between quantities; graph equations on coordinate axes with labels and scales.	**Y**
A-CED.3	3. Represent constraints by equations or inequalities, and by systems of equations and/or inequalities, and interpret solutions as viable or non-viable options in a modeling context. *For example, represent inequalities describing nutritional and cost constraints on combinations of different foods.*	**Y**
A-CED.4	4. Rearrange formulas to highlight a quantity of interest, using the same reasoning as in solving equations. *For example, rearrange Ohm's law* $V = IR$ *to highlight resistance* R.	**Y**

Code	Standard	SAT
Reasoning with Equations and Inequalities [A-REI]		
Understand solving equations as a process of reasoning and explain the reasoning.		
A-REI.1	1. Explain each step in solving a simple equation as following from the equality of numbers asserted at the previous step, starting from the assumption that the original equation has a solution. Construct a viable argument to justify a solution method.	**Y**
A-REI.2	2. Solve simple rational and radical equations in one variable, and give examples showing how extraneous solutions may arise.	**Y**
Solve equations and inequalities in one variable.		
A-REI.3	3. Solve linear equations and inequalities in one variable, including equations with coefficients represented by letters.	**Y**
A-REI.4	4. Solve quadratic equations in one variable.	—
A-REI.4a	a. Use the method of completing the square to transform any quadratic equation in x into an equation of the form $(x-p)2=q$ that has the same solutions. Derive the quadratic formula from this form.	**Y**
A-REI.4b	b. Solve quadratic equations by inspection (e.g., for $x2=49$), taking square roots, completing the square, the quadratic formula, and factoring, as appropriate to the initial form of the equation. Recognize when the quadratic formula gives complex solutions and write them as $a \pm bi$ for real numbers a and b.	**Y**
Solve systems of equations.		
A-REI.5	5. Prove that, given a system of two equations in two variables, replacing one equation by the sum of that equation and a multiple of the other produces a system with the same solutions.	**P**
A-REI.6	6. Solve systems of linear equations exactly and approximately (e.g., with graphs), focusing on pairs of linear equations in two variables.	**Y**
A-REI.7	7. Solve a simple system consisting of a linear equation and a quadratic equation in two variables algebraically and graphically. *For example, find the points of intersection between the line* $y=-3x$ *and the circle* $x^2+y^2=3$.	**Y**
A-REI.8	8. (+) Represent a system of linear equations as a single matrix equation in a vector variable.	**N**

(Continued)

Code	Standard	SAT
A-REI.9	9. (+) Find the inverse of a matrix if it exists and use it to solve systems of linear equations (using technology for matrices of dimension 3×3 or greater).	**N**
Represent and solve equations and inequalities graphically.		
A-REI.10	10. Understand that the graph of an equation in two variables is the set of all its solutions plotted in the coordinate plane, often forming a curve (which could be a line).	**Y**
A-REI.11	11. Explain why the *x*-coordinates of the points where the graphs of the equations $y = f(x)$ and $y = g(x)$ intersect are the solutions of the equation $f(x) = g(x)$; find the solutions approximately, e.g., using technology to graph the functions, make tables of values, or find successive approximations. Include cases where $f(x)$ and/or $g(x)$ are linear, polynomial, rational, absolute value, exponential, and logarithmic functions.*	**P**
A-REI.12	12. Graph the solutions to a linear inequality in two variables as a half-plane (excluding the boundary in the case of a strict inequality), and graph the solution set to a system of linear inequalities in two variables as the intersection of the corresponding half-planes.	**Y**

* indicates Modeling standard (+) indicates standard beyond College and Career Ready

Code	Standard	SAT
Interpreting Functions [F-IF]		
Understand the concept of a function and use function notation.		
F-IF.1	1. Understand that a function from one set (called the domain) to another set (called the range) assigns to each element of the domain exactly one element of the range. If f is a function and x is an element of its domain, then $f(x)$ denotes the output of f corresponding to the input x. The graph of f is the graph of the equation $y = f(x)$.	**Y**
F-IF.2	2. Use function notation, evaluate functions for inputs in their domains, and interpret statements that use function notation in terms of a context.	**Y**
F-IF.3	3. Recognize that sequences are functions, sometimes defined recursively, whose domain is a subset of the integers. For example, the Fibonacci sequence is defined recursively by $f(0) = f(1) = 1$, $f(n + 1) = f(n) + f(n - 1)$ for $n \geq 1$.	**Y**
Interpret functions that arise in applications in terms of the context.		
F-IF.4	4. For a function that models a relationship between two quantities, interpret key features of graphs and tables in terms of the quantities, and sketch graphs showing key features given a verbal description of the relationship. *Key features include intercepts; intervals where the function is increasing, decreasing, positive, or negative; relative maximums and minimums; symmetries; end behavior; and periodicity.**	**Y**
F-IF.5	5. Relate the domain of a function to its graph and, where applicable, to the quantitative relationship it describes. *For example, if the function h(n) gives the number of person-hours it takes to assemble n engines in a factory, then the positive integers would be an appropriate domain for the function.**	**Y**
F-IF.6	6. Calculate and interpret the average rate of change of a function (presented symbolically or as a table) over a specified interval. Estimate the rate of change from a graph.*	**Y**
Analyze functions using different representations.		
F-IF.7	7. Graph functions expressed symbolically and show key features of the graph, by hand in simple cases and using technology for more complicated cases.*	—

(Continued)

Code	Standard	SAT
F-IF.7a	a. Graph linear and quadratic functions and show intercepts, maxima, and minima.*	**Y**
F-IF.7b	b. Graph square root, cube root, and piecewise-defined functions, including step functions and absolute value functions.*	**Y**
F-IF.7c	c. Graph polynomial functions, identifying zeros when suitable factorizations are available, and showing end behavior.*	**Y**
F-IF.7d	d. (+) Graph rational functions, identifying zeros and asymptotes when suitable factorizations are available, and showing end behavior.*	**Y**
F-IF.7e	e. Graph exponential and logarithmic functions, showing intercepts and end behavior, and trigonometric functions, showing period, midline, and amplitude.*	**Y**
F-IF.8	8. Write a function defined by an expression in different but equivalent forms to reveal and explain different properties of the function.	—
F-IF.8a	a. Use the process of factoring and completing the square in a quadratic function to show zeros, extreme values, and symmetry of the graph, and interpret these in terms of a context.	**Y**
F-IF.8b	b. Use the properties of exponents to interpret expressions for exponential functions. *For example, identify percent rate of change in functions such as y = (1.02)t, y = (0.97)t, y = (1.01)12t, and y = (1.2)t/10, and classify them as representing exponential growth or decay.*	**Y**
F-IF.9	9. Compare properties of two functions each represented in a different way (algebraically, graphically, numerically in tables, or by verbal descriptions). *For example, given a graph of one quadratic function and an algebraic expression for another, say which has the larger maximum.*	**Y**

Building Functions [F-BF]

Build a function that models a relationship between two quantities.

F-BF.1	1. Write a function that describes a relationship between two quantities.*	—
F-BF.1a	a. Determine an explicit expression, a recursive process, or steps for calculation from a context.*	**Y**

Code	Standard	SAT
F-BF.1b	b. Combine standard function types using arithmetic operations. *For example, build a function that models the temperature of a cooling body by adding a constant function to a decaying exponential, and relate these functions to the model.**	**Y**
F-BF.1c	c. (+) Compose *functions. For example, if T(y) is the temperature in the atmosphere as a function of height, and h(t) is the height of a weather balloon as a function of time, then T(h(t)) is the temperature at the location of the weather balloon as a function of time.**	**Y**
F-BF.2	2. Write arithmetic and geometric sequences both recursively and with an explicit formula, use them to model situations, and translate between the two forms.*	**Y**
Build new functions from existing functions.		
F-BF.3	3. Identify the effect on the graph of replacing $f(x)$ by $f(x) + k$, $k\,f(x)$, $f(kx)$, and $f(x + k)$ for specific values of k (both positive and negative); find the value of k given the graphs. Experiment with cases and illustrate an explanation of the effects on the graph using technology. *Include recognizing even and odd functions from their graphs and algebraic expressions for them.*	**Y**
F-BF.4	4. Find inverse functions.	—
F-BF.4a	a. Solve an equation of the form $f(x) = c$ for a simple function f that has an inverse and write an expression for the inverse. *For example,* $f(x) = 2x3$ *or* $f(x) = (x + 1)/(x - 1)$ *for* $x \neq 1$.	**Y**
F-BF.4b	b. (+) Verify by composition that one function is the inverse of another.	**Y**
F-BF.4c	c. (+) Read values of an inverse function from a graph or a table, given that the function has an inverse.	**Y**
F-BF.4d	d. (+) Produce an invertible function from a non-invertible function by restricting the domain.	**Y**
F-BF.5	5. (+) Understand the inverse relationship between exponents and logarithms and use this relationship to solve problems involving logarithms and exponents.	**Y**

(*Continued*)

Code	Standard	SAT
Linear, Quadratic, and Exponential Models [F-LE]		
Construct and compare linear, quadratic, and exponential models and solve problems.		
F-LE.1	1. Distinguish between situations that can be modeled with linear functions and with exponential functions.*	—
F-LE.1a	a. Prove that linear functions grow by equal differences over equal intervals, and that exponential functions grow by equal factors over equal intervals.*	**Y**
F-LE.1b	b. Recognize situations in which one quantity changes at a constant rate per unit interval relative to another.*	**Y**
F-LE.1c	c. Recognize situations in which a quantity grows or decays by a constant percent rate per unit interval relative to another.*	**Y**
F-LE.2	2. Construct linear and exponential functions, including arithmetic and geometric sequences, given a graph, a description of a relationship, or two input–output pairs (include reading these from a table).*	**Y**
F-LE.3	3. Observe using graphs and tables that a quantity increasing exponentially eventually exceeds a quantity increasing linearly, quadratically, or (more generally) as a polynomial function.*	**Y**
F-LE.4	4. For exponential models, express as a logarithm the solution to $abct = d$ where a, c, and d are numbers and the base b is 2, 10, or e; evaluate the logarithm using technology.*	**N**
Interpret expressions for functions in terms of the situation they model.		
F-LE.5	5. Interpret the parameters in a linear or exponential function in terms of a context.*	**Y**
Trigonometric Functions [F-TF]		
Extend the domain of trigonometric functions using the unit circle.		
F-TF.1	1. Understand radian measure of an angle as the length of the arc on the unit circle subtended by the angle.	**Y**

Code	Standard	SAT
F-TF.2	2. Explain how the unit circle in the coordinate plane enables the extension of trigonometric functions to all real numbers, interpreted as radian measures of angles traversed counterclockwise around the unit circle.	**Y**
F-TF.3	3. (+) Use special triangles to determine geometrically the values of sine, cosine, tangent for $\pi/3$, $\pi/4$, and $\pi/6$, and use the unit circle to express the values of sine, cosine, and tangent for $\pi - x$, $\pi + x$, and $2\pi - x$ in terms of their values for x, where x is any real number.	**Y**
F-TF.4	4. (+) Use the unit circle to explain symmetry (odd and even) and periodicity of trigonometric functions.	**Y**
Model periodic phenomena with trigonometric functions.		
F-TF.5	5. Choose trigonometric functions to model periodic phenomena with specified amplitude, frequency, and midline.*	**N**
F-TF.6	6. (+) Understand that restricting a trigonometric function to a domain on which it is always increasing or always decreasing allows its inverse to be constructed.	**N**
F-TF.7	7. (+) Use inverse functions to solve trigonometric equations that arise in modeling contexts; evaluate the solutions using technology; and interpret them in terms of the context.*	**N**
Prove and apply trigonometric identities.		
F-TF.8	8. Prove the Pythagorean identity $\sin2(\theta) + \cos2(\theta) = 1$ and use it to find $\sin(\theta)$, $\cos(\theta)$, or $\tan(\theta)$ given $\sin(\theta)$, $\cos(\theta)$, or $\tan(\theta)$ and the quadrant.	**N**
F-TF.9	9. (+) Prove the addition and subtraction formulas for sine, cosine, and tangent and use them to solve problems.	**N**

* indicates modeling standard (+) indicates standard beyond college and career readiness

Code	Standard	SAT
Congruence [G-CO]		
Experiment with transformations in the plane.		
G-CO.1	1. Know precise definitions of angle, circle, perpendicular line, parallel line, and line segment, based on the undefined notions of point, line, distance along a line, and distance around a circular arc.	**Y**
G-CO.2	2. Represent transformations in the plane using, e.g., transparencies and geometry software; describe transformations as functions that take points in the plane as inputs and give other points as outputs. Compare transformations that preserve distance and angle to those that do not (e.g., translation versus horizontal stretch).	**Y**
G-CO.3	3. Given a rectangle, parallelogram, trapezoid, or regular polygon, describe the rotations and reflections that carry it onto itself.	**Y**
G-CO.4	4. Develop definitions of rotations, reflections, and translations in terms of angles, circles, perpendicular lines, parallel lines, and line segments.	**Y**
G-CO.5	5. Given a geometric figure and a rotation, reflection, or translation, draw the transformed figure using, e.g., graph paper, tracing paper, or geometry software. Specify a sequence of transformations that will carry a given figure onto another.	**Y**
Understand congruence in terms of rigid motions.		
G-CO.6	6. Use geometric descriptions of rigid motions to transform figures and to predict the effect of a given rigid motion on a given figure; given two figures, use the definition of congruence in terms of rigid motions to decide if they are congruent.	**Y**
G-CO.7	7. Use the definition of congruence in terms of rigid motions to show that two triangles are congruent if and only if corresponding pairs of sides and corresponding pairs of angles are congruent.	**Y**

Code	Standard	SAT
G-CO.8	8. Explain how the criteria for triangle congruence (ASA, SAS, and SSS) follow from the definition of congruence in terms of rigid motions.	**Y**
Prove geometric theorems.		
G-CO.9	9. Prove theorems about lines and angles. *Theorems include vertical angles are congruent; when a transversal crosses parallel lines, alternate interior angles are congruent and corresponding angles are congruent; points on a perpendicular bisector of a line segment are exactly those equidistant from the segment's endpoints.*	**P**
G-CO.10	10. Prove theorems about triangles. *Theorems include measures of interior angles of a triangle sum to 180°; base angles of isosceles triangles are congruent; the segment joining midpoints of two sides of a triangle is parallel to the third side and half the length; the medians of a triangle meet at a point.*	**P**
G-CO.11	11. Prove theorems about parallelograms. *Theorems include opposite sides are congruent; opposite angles are congruent; the diagonals of a parallelogram bisect each other; and, conversely, rectangles are parallelograms with congruent diagonals.*	**P**
Make geometric constructions.		
G-CO.12	12. Make formal geometric constructions with a variety of tools and methods (compass and straightedge, string, reflective devices, paper folding, dynamic geometric software, etc.). *Copying a segment; copying an angle; bisecting a segment; bisecting an angle; constructing perpendicular lines, including the perpendicular bisector of a line segment; and constructing a line parallel to a given line through a point not on the line.*	**Y**
G-CO.13	13. Construct an equilateral triangle, a square, and a regular hexagon inscribed in a circle.	**Y**
Similarity, Right Triangles, and Trigonometry [G-SRT]		
Understand similarity in terms of similarity transformations.		
G-SRT.1	1. Verify experimentally the properties of dilations given by a center and a scale factor:	—

(Continued)

Code	Standard	SAT
G-SRT.1a	a. A dilation takes a line not passing through the center of the dilation to a parallel line, and leaves a line passing through the center unchanged.	**Y**
G-SRT.1b	b. The dilation of a line segment is longer or shorter in the ratio given by the scale factor.	**Y**
G-SRT.2	2. Given two figures, use the definition of similarity in terms of similarity transformations to decide if they are similar; explain using similarity transformations the meaning of similarity for triangles as the equality of all corresponding pairs of angles and the proportionality of all corresponding pairs of sides.	**Y**
G-SRT.3	3. Use the properties of similarity transformations to establish the Angle-Angle (AA) criterion for two triangles to be similar.	**Y**
Prove theorems involving similarity.		
G-SRT.4	4. Prove theorems about triangles. *Theorems include a line parallel to one side of a triangle, divides the other two proportionally, and, conversely, the Pythagorean Theorem proved using triangle similarity.*	**P**
G-SRT.5	5. Use congruence and similarity criteria for triangles to solve problems and to prove relationships in geometric figures.	**Y**
Define trigonometric ratios and solve problems involving right triangles.		
G-SRT.6	6. Understand that by similarity, side ratios in right triangles are properties of the angles in the triangle, leading to definitions of trigonometric ratios for acute angles.	**Y**
G-SRT.7	7. Explain and use the relationship between the sine and cosine of complementary angles.	**Y**
G-SRT.8	8. Use trigonometric ratios and the Pythagorean Theorem to solve right triangles in applied problems.*	**Y**
Apply trigonometry to general triangles.		
G-SRT.9	9. (+) Derive the formula $A = \frac{1}{2}ab \sin(C)$ for the area of a triangle by drawing an auxiliary line from a vertex perpendicular to the opposite side.	**N**
G-SRT.10	10. (+) Prove the Laws of Sines and Cosines and use them to solve problems.	**N**

Code	Standard	SAT
G-SRT.11	11. (+) Understand and apply the Law of Sines and the Law of Cosines to find unknown measurements in right and nonright triangles (e.g., surveying problems, resultant forces).	**N**
Circles [G-C]		
Understand and apply theorems about circles.		
G-C.1	1. Prove that all circles are similar.	**P**
G-C.2	2. Identify and describe relationships among inscribed angles, radii, and chords. *Include the relationship between central, inscribed, and circumscribed angles; inscribed angles on a diameter are right angles; the radius of a circle is perpendicular to the tangent where the radius intersects the circle.*	**Y**
G-C.3	3. Construct the inscribed and circumscribed circles of a triangle, and prove properties of angles for a quadrilateral inscribed in a circle.	**P**
G-C.4	4. (+) Construct a tangent line from a point outside a given circle to the circle.	**P**
Find arc lengths and areas of sectors of circles.		
G-C.5	5. Derive using similarity the fact that the length of the arc intercepted by an angle is proportional to the radius, and define the radian measure of the angle as the constant of proportionality; derive the formula for the area of a sector.	**Y**
Expressing Geometric Properties with Equations G-GPE		
Translate between the geometric description and the equation for a conic section.		
G-GPE.1	1. Derive the equation of a circle of given center and radius using the Pythagorean Theorem; complete the square to find the center and radius of a circle given by an equation.	**Y**
G-GPE.2	2. Derive the equation of a parabola given a focus and directrix.	**Y**
G-GPE.3	3. (+) Derive the equations of ellipses and hyperbolas given the foci, using the fact that the sum or difference of distances from the foci is constant.	**Y**

(Continued)

Code	Standard	SAT
Use coordinates to prove simple geometric theorems algebraically.		
G-GPE.4	4. Use coordinates to prove simple geometric theorems algebraically. For example, prove or disprove that a figure defined by four given points in the coordinate plane is a rectangle; prove or disprove that the point $(1,\sqrt{3})$ lies on the circle centered at the origin and containing the point (0, 2).	**Y**
G-GPE.5	5. Prove the slope criteria for parallel and perpendicular lines and use them to solve geometric problems (e.g., find the equation of a line parallel or perpendicular to a given line that passes through a given point).	**Y**
G-GPE.6	6. Find the point on a directed line segment between two given points that partitions the segment in a given ratio.	**Y**
G-GPE.7	7. Use coordinates to compute perimeters of polygons and areas of triangles and rectangles, e.g., using the distance formula.*	**Y**
Geometric Measurement and Dimension [G-GMD]		
Explain volume formulas and use them to solve problems.		
G-GMD.1	1. Give an informal argument for the formulas for the circumference of a circle, area of a circle, volume of a cylinder, pyramid, and cone. *Use dissection arguments, Cavalieri's principle, and informal limit arguments.*	**P**
G-GMD.2	2. (+) Give an informal argument using Cavalieri's principle for the formulas for the volume of a sphere and other solid figures.	**P**
G-GMD.3	3. Use volume formulas for cylinders, pyramids, cones, and spheres to solve problems.*	**Y**
Visualize relationships between two-dimensional and three-dimensional objects.		
G-GMD.4	4. Identify the shapes of two-dimensional cross-sections of three-dimensional objects, and identify three-dimensional objects generated by rotations of two-dimensional objects.	**Y**

Code	Standard	SAT
Modeling with Geometry [G-MG]		
Apply geometric concepts in modeling situations.		
G-MG.1	1. Use geometric shapes, their measures, and their properties to describe objects (e.g., modeling a tree trunk or a human torso as a cylinder).*	**Y**
G-MG.2	2. Apply concepts of density based on area and volume in modeling situations (e.g., persons per square mile, BTUs per cubic foot.*	**Y**
G-MG.3	3. Apply geometric methods to solve design problems (e.g., designing an object or structure to satisfy physical constraints or minimize cost; working with typographic grid systems based on ratios).*	**Y**

* indicates modeling standard (+) indicates standard beyond college and career readiness

Code	Standard	SAT
Interpreting Categorical and Quantitative Data [S-ID]		
Summarize, represent, and interpret data on a single count or measurement variable.		
S-ID.1	1. Represent data with plots on the real number line (dot plots, histograms, and box plots).*	**Y**
S-ID.2	2. Use statistics appropriate to the shape of the data distribution to compare center (median, mean) and spread (interquartile range, standard deviation) of two or more different data sets.*	**Y**
S-ID.3	3. Interpret differences in shape, center, and spread in the context of the data sets, accounting for possible effects of extreme data points (outliers).*	**Y**
S-ID.4	4. Use the mean and standard deviation of a data set to fit it to a normal distribution and to estimate population percentages. Recognize that there are data sets for which such a procedure is not appropriate. Use calculators, spreadsheets, and tables to estimate areas under the normal curve.*	**Y**
Summarize, represent, and interpret data on two categorical and quantitative variables.		
S-ID.5	5. Summarize categorical data for two categories in two-way frequency tables. Interpret relative frequencies in the context of the data (including joint, marginal, and conditional relative frequencies). Recognize possible associations and trends in the data.*	**Y**
S-ID.6	6. Represent data on two quantitative variables on a scatter plot, and describe how the variables are related.*	—
S-ID.6a	a. Fit a function to the data; use functions fitted to data to solve problems in the context of the data. Use given functions or choose a function suggested by the context. Emphasize linear, quadratic, and exponential models.*	**Y**
S-ID.6b	b. Informally assess the fit of a function by plotting and analyzing residuals.*	**Y**
S-ID.6c	c. Fit a linear function for a scatter plot that suggests a linear association.*	**Y**

Code	Standard	SAT
Interpret linear models.		
S-ID.7	7. Interpret the slope (rate of change) and the intercept (constant term) of a linear model in the context of the data.*	Y
S-ID.8	8. Compute (using technology) and interpret the correlation coefficient of a linear fit.*	N
S-ID.9	9. Distinguish between correlation and causation.*	Y
Making Inferences and Justifying Conclusions [S-IC]		
Understand and evaluate random processes underlying statistical experiments.		
S-IC.1	1. Understand statistics as a process for making inferences to be made about population parameters based on a random sample from that population.*	Y
S-IC.2	2. Decide if a specified model is consistent with results from a given data-generating process, e.g., using simulation. For example, a model says a spinning coin falls heads up with probability 0.5. Would a result of 5 tails in a row cause you to question the model?*	Y
Make inferences and justify conclusions from sample surveys, experiments, and observational studies.		
S-IC.3	3. Recognize the purposes of and differences among sample surveys, experiments, and observational studies; explain how randomization relates to each.*	Y
S-IC.4	4. Use data from a sample survey to estimate a population mean or proportion; develop a margin of error through the use of simulation models for random sampling.*	Y
S-IC.5	5. Use data from a randomized experiment to compare two treatments; use simulations to decide if differences between parameters are significant.*	Y
S-IC.6	6. Evaluate reports based on data.*	Y
Conditional Probability and the Rules of Probability [S-CP]		
Understand independence and conditional probability and use them to interpret data.		
S-CP.1	1. Describe events as subsets of a sample space (the set of outcomes) using characteristics (or categories) of the outcomes, or as unions, intersections, or complements of other events ("or", "and", "not").*	Y

(Continued)

Code	Standard	SAT
S-CP.2	2. Understand that two events *A* and *B* are independent if the probability of *A* and *B* occurring together is the product of their probabilities, and use this characterization to determine if they are independent.*	**Y**
S-CP.3	3. Understand the conditional probability of *A* given *B* as *P*(*A* and *B*)/*P*(*B*), and interpret independence of *A* and *B* as saying that the conditional probability of *A* given *B* is the same as the probability of *A*, and the conditional probability of *B* given *A* is the same as the probability of *B*.*	**Y**
S-CP.4	4. Construct and interpret two-way frequency tables of data when two categories are associated with each object being classified. Use the two-way table as a sample space to decide if events are independent and to approximate conditional probabilities. *For example, collect data from a random sample of students in your school on their favorite subject among math, science, and English. Estimate the probability that a randomly selected student from your school will favor science given that the student is in tenth grade. Do the same for other subjects and compare the results.**	**Y**
S-CP.5	5. Recognize and explain the concepts of conditional probability and independence in everyday language and everyday situations. *For example, compare the chance of having lung cancer if you are a smoker with the chance of being a smoker if you have lung cancer.**	**Y**
Use the rules of probability to compute probabilities of compound events in a uniform probability model.		
S-CP.6	6. Find the conditional probability of *A* given *B* as the fraction of *B*'s outcomes that also belong to *A*, and interpret the answer in terms of the model.*	**Y**
S-CP.7	7. Apply the Addition Rule, $P(A \text{ or } B) = P(A) + P(B) - P(A \text{ and } B)$, and interpret the answer in terms of the model.*	**Y**
S-CP.8	8. (+) Apply the general Multiplication Rule in a uniform probability model, $P(A \text{ and } B) = P(A)P(B\|A) = P(B)P(A\|B)$, and interpret the answer in terms of the model.*	**Y**

Code	Standard	SAT
S-CP.9	9. (+) Use permutations and combinations to compute probabilities of compound events and solve problems.*	**Y**
Using Probability to Make Decisions [S-MD]		
Calculate expected values and use them to solve problems.		
S-MD.1	1. (+) Define a random variable for a quantity of interest by assigning a numerical value to each event in a sample space; graph the corresponding probability distribution using the same graphical displays as for data distributions.*	**Y**
S-MD.2	2. (+) Calculate the expected value of a random variable; interpret it as the mean of the probability distribution.*	**N**
S-MD.3	3. (+) Develop a probability distribution for a random variable defined for a sample space in which theoretical probabilities can be calculated; find the expected value. *For example, find the theoretical probability distribution for the number of correct answers obtained by guessing on all five questions of a multiple-choice test where each question has four choices, and find the expected grade under various grading schemes.**	**Y**
S-MD.4	4. (+) Develop a probability distribution for a random variable defined for a sample space in which probabilities are assigned empirically; find the expected value. *For example, find a current data distribution on the number of TV sets per household in the United States, and calculate the expected number of sets per household. How many TV sets would you expect to find in 100 randomly selected households?**	**Y**
Use probability to evaluate outcomes of decisions.		
S-MD.5	5. (+) Weigh the possible outcomes of a decision by assigning probabilities to payoff values and finding expected values.*	—
S-MD.5a	a. (+) Find the expected payoff for a game of chance. For example, find the expected winnings from a state lottery ticket or a game at a fast-food restaurant.*	**Y**
S-MD.5b	b. (+) Evaluate and compare strategies on the basis of expected values. For example, compare a high-deductible versus a low-deductible automobile insurance policy using various, but reasonable, chances of having a minor or a major accident.*	**Y**

(Continued)

Code	Standard	SAT
S-MD.6	6. (+) Use probabilities to make fair decisions (e.g., drawing by lots, using a random number generator).*	**Y**
S-MD.7	7. (+) Analyze decisions and strategies using probability concepts (e.g., product testing, medical testing, pulling a hockey goalie at the end of a game).*	**Y**

* indicates modeling standard (+) indicates standard beyond college and career readiness

Additional Resources

A-List

Main website: www.alisteducation.com
Bookstore: www.alisteducation.com/bookstore

The SAT

Main website: http://sat.collegeboard.org/home
Test specifications for the redesigned SAT: https://collegereadiness.collegeboard.org/pdf/test-specifications-redesigned-sat.pdf
Free Practice tests: https://collegereadiness.collegeboard.org/sat/practice/full-length-practice-tests
The same practice tests are also available in *The Official SAT Study Guide,* College Board, 2016.
Additional sample questions: https://collegereadiness.collegeboard.org/sample-questions
College Board Guide to Implementing the Redesigned SAT: https://collegereadiness.collegeboard.org/pdf/college-board-guide-implementing-redesigned-sat-installment-2.pdf
SAT-Common Core alignment study (pdf): http://professionals.collegeboard.com/profdownload/pdf/10b_2901_Comm_Core_Report_Complete_WEB_101117.pdf
Additional practice at Khan Academy (requires free login): www.khanacademy.org/

The Common Core State Standards

Main website: www.corestandards.org/
The Standards (available to read on the web or as pdf downloads): www.corestandards.org/the-standards

For Product Safety Concerns and Information please contact our EU
representative GPSR@taylorandfrancis.com
Taylor & Francis Verlag GmbH, Kaufingerstraße 24, 80331 München, Germany

www.ingramcontent.com/pod-product-compliance
Ingram Content Group UK Ltd.
Pitfield, Milton Keynes, MK11 3LW, UK
UKHW052053230926
475702UK00025B/1153